"In this wonderfully creative and imaginative account, a proton tells the story of both science and faith from beginning to end. This new and refreshing perspective keeps the reader engaged from start to finish—the story of the universe in a style we encounter for the very first time. A gripping read and recommended for readers of any age."

Denis Alexander, emeritus director of the Faraday Institute for Science and Religion, Cambridge University

"This is a fantastic book, one that I would have liked to have written myself. Great that it is here now. Highly recommended!"

Heino Falcke, professor of astrophysics at Radboud University, the Netherlands, and author of *Light in the Darkness: Black Holes, the Universe, and Us*

"I have always dreamed of traversing the entire universe. But the authors of this fascinating book go a step further—they traverse the entirety of time, all 13.7 billion years of it! From a perspective firmly grounded in love for the Creator and respect for his creatures, they have given us an imaginative tour de force."

Matthew Levering, James N. Jr. and Mary D. Perry Chair of Theology at Mundelein Seminary

"This is a remarkable book combining good science, rich and childlike imagination, and deeply grounded creational theology. The story of the universe told by a proton! Who comes up with that? It's great for children, and adults won't be able to resist it!"

W. Ross Hastings, Sangwoo Youtong Chee Professor of Theology at Regent College

"*Dawn* is a unique and engaging book arising from a conversation between scientists, philosophers, and theologians . . . they found a grand story to tell. Starting from the Big Bang, a few protons are our guides, proverbial flies on a wall. Both a work of fiction and nonfiction, this story is a fresh invitation to the great and grand story of who we are and who is God."

S. Joshua Swamidass, associate professor of laboratory and genomic medicine at Washington University in St Louis

DAWN

A PROTON'S TALE OF ALL THAT CAME TO BE

CEES DEKKER, CORIEN ORANJE, AND
GIJSBERT VAN DEN BRINK

TRANSLATED BY HARRY COOK
AFTERWORD BY DEBORAH HAARSMA

An imprint of InterVarsity Press
Downers Grove, Illinois

InterVarsity Press
P.O. Box 1400 | Downers Grove, IL 60515-1426
ivpress.com | email@ivpress.com

Originally published by ArkMedia, a division of Jongbloed Uitgeverij bv, as *Oer: Het grote verhaal van nul tot nu* by Corien Oranje, Gijsbert van den Brink, and Cees Dekker.

InterVarsity Press® is the publishing division of InterVarsity Christian Fellowship/USA®.
For more information, visit intervarsity.org.

Cover design and image composite: David Fassett
Interior design: Jeanna Wiggins

ISBN 978-1-5140-0566-8 (print) | ISBN 978-1-5140-0567-5 (digital)

Printed in the United States of America ♾

Library of Congress Cataloging-in-Publication Data
A catalog record for this book is available from the Library of Congress.

29 28 27 25 24 23 22 21 20 | 8 7 6 5 4 3 2 1

CONTENTS

PROLOGUE

WHAT AN UNBELIEVABLE JOURNEY. What a dizzying rollercoaster ride. An adventure that began almost fourteen billion years ago, one that so often threatened to fail. It's truly a miracle I'm still here. Despite everything, I wouldn't have wanted to miss one second of it. And the best is yet to come.

"Make sure it gets written down," said my friends.

"For whom?" I asked. "You were right there too, for all of it!"

"For the human beings," they said. "Do it for them, since it's not only our story, it's their story as well. They have to know it, and they're pretty new on the scene. We should tell them what we've experienced."

I wasn't immediately taken with the idea. "*Homo sapiens*? They're so limited. They wouldn't be able to grasp it."

"It doesn't matter," they said. "Use their language, use words they can understand. Just try it, Pro. Just so they can get a little bit of an idea."

Okay then, I'll give it a try.

—Proton

1

BEGINNINGS

I WAS BORN DURING a messy runaway frenzy. Collisions. Chaos. Flying objects.

Perhaps you should compare it to a fireworks factory kindled by a spark. Explosions that followed each other at furious speed. The heat. The pressure. You could call it a cyclone, a raging EF5 tornado that took nothing and no one into account, and in which everything and everyone was flung away and destroyed.

"Behind you," someone yelled. I couldn't escape; I was pushed and thrown in every direction. To my consternation, I noticed that another newborn proton that had been floating beside me a moment ago, was blown away, far out of reach. There went another, and another. Everything spun, turned, dove, and crossed paths at lightning speed.

"Do you know what's going on?" a voice called out.

"I have no idea," I answered. I whirled helplessly while projectiles flew past me on all sides. Around me, time and space exploded. There was nothing I could do to protect myself, to take myself out of the line of fire. A safe place was nowhere to be found.

• • •

I must have lost consciousness at some point. When I woke up, I heard voices. It sounded like they were not far from where I was, but I couldn't see anything.

"How terrible," said the one voice.

"That was disturbing indeed," said the other voice.

"Disturbing?" The first voice sounded indignant. "It was way more than disturbing. It was a disaster. A slaughter! No one is left. Everything is destroyed! Everything and everyone."

I did not understand much of this conversation, but it was clear that the speakers thought they were the only ones left.

"And what about you and me?" the first voice asked.

"We're the only ones!"

"Not at all."

"Hello," I called out cautiously.

The two voices fell silent.

"Hello," I said again. "Is there anybody out there?"

"Who are you?" the first voice asked.

"I'm Proton," I replied.

"I told you!" the other voice said triumphantly. "Ha! I told you we were not alone. Hey, Proton, my name is Kalon. That faint-hearted friend of mine is Achaton."

"What's happening?" I asked. "What's the problem?"

"He is new here," said the voice that had introduced himself as Kalon. "We have to tell him about it."

The two strangers told me about the past, about the beginning of time and space many eons ago. They told me about someone they called the Creator and to whom I apparently owed my existence. It was a bizarre story, and I found it hard to believe.

"Previously, nothing existed," said Kalon. "Nothing at all. No matter. No energy. No time. No space."

"There was the Creator," said Achaton.

Kalon laughed. "Well, of course the Creator was there. He was always there. He thought of a plan to make something; something grand, something exceptional, something . . ."

" . . . spectacular!" cried Achaton. "Honestly, Proton, you wouldn't believe this!"

"The Creator made an egg," Kalon continued. "A minuscule egg."

I didn't really have any idea of what an egg was, but it seemed smart to me to not interrupt Kalon—I didn't want to appear stupid.

"A seed," said Kalon, who fortunately did not catch on to my ignorance. "A speck. Smaller than a speck. Incredibly heavy and unbelievably hot."

"How hot?" I asked.

"A million times a trillion times a trillion degrees Celsius," Achaton said.

"Much hotter even," said Kalon. "And that seed contained everything! All building materials. All energy. All forces and laws of nature. Everything the Creator needed. He put all his wisdom into it. His greatness."

"And then it happened!" proclaimed Achaton, who could hardly contain his excitement. "He said something, didn't he, Kalon? 'There must be light.' He said something like that."

"That's what he said," agreed Kalon. "The Creator spoke, and then something happened that only he could have invented. All powers that the Creator had concentrated into that one speck erupted, spattering apart with a speed and a heat we cannot imagine anymore. A burning hot universe expanded itself in all directions. All of a sudden, everything was there and time had begun."

It was all a little too fast for me. I was still reflecting, baffled by the mysterious sentence that the one called "the Creator" had supposedly spoken. "What is light?" I asked.

Kalon was silent for a moment. "*That* we do not know," he answered. "It's a mystery. There are many mysteries. I don't understand everything either, Proton. Maybe it's something that was there in the very beginning. Something that was necessary to start the whole process. Or maybe there will come a moment when we will think, Aha, *that* is light."

Suddenly, there was a head-on collision nearby. "Whoa!" "What was that?" "That was a very close call!" It made me realize how precarious our situation was, but Kalon was not to be deterred. "Now, where was I?" he asked.

"The second eon," said Achaton. "You were going to tell us about the second eon."

"Ah, yes! A very special period in history."

"What period, roughly, are we talking about?" I asked.

"Oh, that is now so long ago. Time had only just begun. It was even before a trillionth of a trillionth of a trillionth of a second had passed, the moment when gravity could finally begin its work. The Creator had adjusted it with the greatest possible accuracy beforehand, and for good reason. The existence of his creation was at stake. If gravity had been just a fraction stronger, the whole universe would have shrunk immediately. A tiny bit weaker and everything would have dispersed and become too thin. Then no particles would have formed whatsoever."

"Then I would not have existed," I said somewhat anxiously.

"Precisely! And that was only the beginning. In the eons that followed . . ."

"Yes, tell him about the quarks," Achaton chipped in enthusiastically.

"The *what*?" I asked. Another one of those words I'd never heard before. Everything was new for me and, to be honest, it was far too much for me to comprehend all at once.

"You don't know what quarks are?" Achaton chuckled. "Unbelievable. You are made *up* of quarks. Without quarks you wouldn't exist."

"Please don't act as if you've always known this," said Kalon. He sounded irritated.

"Quarks are just building blocks, Proton. The building blocks that we consist of. Achaton, you, me, and the rest of us. Space expanded, the temperature dropped, and this caused particles to form from the energy. Electrons. Quarks. And antiquarks, of course."

Ah okay, I sighed in relief. At least this was something I could follow, sort of.

• • •

Kalon told me about the two eons that followed, in which the other forces of nature acquired their roles. The Creator came up with four fundamental forces: in addition to the gravitational force there were the strong and weak nuclear forces and the electromagnetic force. Four giants, each with a unique set of tasks. All four had to be set very precisely, and it was a tough call on whether the newborn universe was going to make it. The smallest deviation would have had disastrous consequences. The likelihood that the universe would destroy itself was billions of times greater than that it would go well.

"That must have been a tense time," I said.

"Tense?" Achaton replied indignantly. "That is an understatement, to put it mildly. It was make it or break it! Four times in a row, the universe had to win the jackpot in a global lottery. If the nuclear forces would have been even a teeny bit stronger, things would have gone seriously awry. And what do you think would've become of the universe if the electromagnetic force would have been a fraction weaker?"

"I don't know," I said. I wished I were more intelligent than I was, so that I could understand exactly what Achaton was telling me. And I wished I had clever answers, but I didn't.

"Then we would not have met you here. Let's keep it at that."

"But the beautiful thing was," said Kalon, "the forces were perfectly attuned to each other. The gravitational force, the electromagnetic force, and the strong and weak nuclear forces—the Creator saw to it that they became friends."

"Well . . ." said Achaton. "Friends . . ."

"Okay, *friends* is perhaps too strong a word. Colleagues then, a team. Partners. They perfectly complement each other, as if they've always worked together. The four of them perform one great dance in honor of the Creator."

Kalon described how the universe suddenly began to expand with greater speed, how it drove on with increasing energy, dark and fast, sizzling hot, extending in all directions.

"Even faster than now?" I asked.

"Ha! There's no comparison. So unbelievably much faster than now. It was beyond extreme. It's impossible to imagine."

Not a second had passed since the Big Bang and another eon had already begun.

"So, small particles had formed from the energy. Countless numbers of particles. For example, the quarks I just mentioned. The ones that make up you and me. The universe was one large construction site full of building materials. There was an unfathomably large number of quarks, all ready for the task the Creator had planned for them—to form matter. Nevertheless, for a time it looked as if they would not be able to fulfill this task because they were destined to be destroyed."

"What!" I said, shocked. I wondered how he knew all this.

"For every building block there was an anti–building block, Proton. I'm sure I mentioned it."

"I don't remember that . . ."

"The antiquarks!" Kalon's voice sounded almost indignant. "I am sure I told you about antiquarks."

"But I have no idea what antiquarks are; how could I?"

"They are anti–building blocks. I just tried to explain it to you. I have no idea why the Creator found it necessary to make them, but he did. And as soon as they would touch a quark, poof! Both would disappear and the only thing that remained was a bit of energy. It was a terrible slaughter. The universe came very, very close to being completely empty. With no more building blocks, just energy that was left. But for every ten billion quarks, one quark was *not* destroyed; what remained was more than enough for what the Creator needed."

"So much waste," Achaton sighed.

"Not waste," Kalon said. "Abundance."

"And then, modern times began," Achaton continued. "The time in which we now live. In which we came into existence. Out of the quarks that were still left over."

"I like to call this the most important time in the history of the universe," Kalon said earnestly.

Achaton laughed. "I have long thought that this is what the Creator intended. This gigantic, vast universe filled with photons, electrons, neutrons, protons. Held together by natural forces that are so perfectly attuned to each other."

"You mean," I said, full of hope, "that we are the big plan of the Creator."

"Yes, that's what I thought," said Kalon. "And so much time has passed since the creation. So many eons. The universe has existed longer than you and I can comprehend. More than a second already. But I have heard it said that the Creator is still extremely active. That he has still bigger plans. That he has decided to continue his work with just a small remainder of us. It's possible. It actually *does* seem like it, you see, as times have become turbulent again. Many, many of us have recently disappeared into nothingness, I'm afraid."

"I even thought that you said all of it would go wrong," said Achaton.

"Not me. That was you, if I remember correctly."

"Well, what would it mean, to 'go wrong'? If the Creator had come up with another plan for us, that would have been fine too. Little by little we've learned that we can trust him. Hang on. Hey! What are you doing?"

"I'm not doing anything! What are *you* doing? You are really squeezing me. Get out of the way!"

"Go away, *you*, get away from me!"

And then it was quiet.

"Kalon?" I called. "Achaton! Is something wrong? Where are you both?"

There was no reply. Again I heard something whiz by me, and I was dragged along into a chaotic, destructive three-dimensional collision course. What was happening, where were Kalon and Achaton? I still have no idea. I never met them again.

2

COSMOS

I DIDN'T HAVE MUCH TIME to mourn the sudden disappearance of Kalon and Achaton. The chaos of flying and violently colliding objects continued at full force. Several times I narrowly escaped a barrage. It dawned on me that there was no guarantee whatsoever that I would escape this mayhem. In fact, I could soon be finished. There was destruction all around me, and it seemed as if protons and neutrons were ready to destroy each other.

"Make way, make way!" I suddenly heard a voice call out. A neutron flew toward me at full speed. At the same time, a neutron and a proton that were tightly attached were approaching me from above. "Get away," they shouted, "get out of the way!"

But it was too late. We couldn't avoid each other. In the nanosecond before the collision, I didn't feel any fear, perhaps just disappointment that this was it; that everything was over before it had even properly begun. That I'd never find out about the plan that Kalon and Achaton had mentioned, the plan of the Creator. That I would never come to know the purpose of this universe.

The blow was heavy. For a very short moment I felt an intense heat that engulfed me and took possession of me. After that there was nothing.

"Is he conscious or not?" I heard someone say. "Is he doing all right?"

"No idea. Hey, you, can you hear me?"

Something bumped into me, and I became vaguely aware of my surroundings. Something had changed. I was no longer alone. I was together with three others: two neutrons and another proton, and we were all firmly attached to each other. That was strange. Until now I'd avoided every proton that I saw, and they had avoided me. It wasn't so much that we disliked each other, but it was as though we were too much alike, like two magnets repelling each other. Now it seemed as if we were stuck together with super glue.

"What happened?" I asked weakly.

"Ah, he has woken up. Finally."

"So, you don't know?" said one of the neutrons. She nudged me. "You don't know what happened? Listen, guys, he doesn't know what happened. Who are you? What's your name?"

I still felt dizzy, but I tried to think straight. Vaguely my name came back into my consciousness. I hesitated. "Proton," I said. "My name is Proton."

"All right, Proton, I'm Ensis, and this is Aris. We're neutrons."

"And I'm Solon," another voice chipped in. "I'm a proton, just like you. Let's call you Pro, okay?"

"Hey, Pro, are you aware that we, the four of us, have just become an atomic nucleus?"

"Sorry? We've become what?"

"The nucleus of an atom. Helium!"

"Just a moment," I said. "I don't understand. Why are we attached to each other? Who *are* you?"

There was some muffled murmuring. "He must have been impacted more by that collision than we realized."

"I hope he hasn't been damaged, has he?"

"He doesn't look injured or anything. It must be the shock."

"Hello?" I called out. "I can hear you. Maybe you could just explain to me what's happening? In a way that I can understand?"

"Just leave it to me," said Solon. "Pro, I understand this is a bit much for you to comprehend. You've been involved in a rather serious collision, just like us. But don't worry, all is well. You have survived the whole event unscathed. However, the collision has had some rather drastic results—we've clicked together into a new unit."

"A new kind of mega-building block," Ensis interrupted, "an atomic nucleus."

"You have to see it as a kind of promotion," continued Solon, "a higher rank."

Aris could not contain her enthusiasm. "The Creator is going to use us to build something!" she cried. "Isn't that tremendous, Pro? We're important! He has invented us! We're needed for the completion of his plan."

"Just a moment, not so fast," I responded. "His plan?"

"Yes, you know it, don't you?"

"I know something about it, of course," I said carefully, "but not everything."

"It seems that the Creator is far from being finished with the universe," explained Ensis. "It must become even bigger, much bigger. A lot of work is being done right now, so that it'll meet all the requirements. For this, building blocks are needed, of course."

"What is the Creator going to build, then?"

"We have heard that he's planning to make beings constructed from atomic nuclei, but then much larger and more complex."

"More complex than we are?" I asked. I was just getting a bit more used to this new situation, which, in my view, was already pretty complex: two protons and two neutrons connected as if they belonged together. But actually, it was ingenious, how we fitted together, like we were created for each other.

"Billions of times more complex, they say. All kinds of beings that will reflect some of the creativity, greatness, and love of the Creator. There's a rumor that they'll be *alive*, but none of us knows what that is, exactly."

I tried to imagine these beings, but my imagination didn't go further than swirling clusters of atoms. "And will they be floating around?"

"Of course not!" Solon sniggered. "They won't be able to manage that, these beings. They'll be very vulnerable and prone to fall apart again quite easily. The Creator is going to build a special home for them. He'll prepare a protected and safe place, somewhere in a far corner of the universe."

"I don't know," I responded, "but to me this looks like a project with enormous risks."

"That is indeed the case," said Ensis. "At any moment in the process all kinds of things could go wrong."

"But the Creator has already taken that into account," said Solon. "It's for good reason that he's chosen to have such a vast surplus of energy and matter. This could go wrong billions of times, and even then, his plan will not be in danger."

"And now what?" I asked.

"Now we wait," said Solon, "until it is time."

I was just going to have to trust this. Trust, however, is not my strongest point.

● ● ●

With the four of us together, we whirled several hundred thousand years in a piping hot soup. That sounds perhaps as if it was a boring time, and, indeed, there wasn't much to do in the universe, but I now know that humans experience time quite differently than we do. In this perhaps, we as protons (and let me not exclude the neutrons) are more like the Creator. In a millionth of a second, an infinite number of things can happen, while billions of years can, for us, pass in the blink of an eye.

At a certain moment, however, I noticed that some things were beginning to change. Until then I had barely been able to distinguish anything within the chaos. It felt as though we were driving through a dense fog in a car with all its headlights turned on, or tumbling through dark clouds in an out-of-control airplane. We couldn't see a thing.

But very slowly the fog seemed to lift a little. Things appeared to get a bit clearer, thanks to flickers that very briefly illuminated things and then disappeared, like the flashing of a faulty fluorescent tube in the world of human beings. This was the work of photons, the carriers of light, who'd been virtually invisible until now. They'd been there from the beginning, but only now could I see their effect: they enabled you to distinguish things in the space around you.

"This is it!" I exclaimed in surprise.

"What?" asked Solon.

"Just have a look!" I said. "This must be light! The light that the Creator mentioned."

I told my nuclear partners, whom I'd begun to see as friends, about when I'd met Kalon and Achaton. How we'd been racking our brains to figure out what light was. This had to be the answer to that puzzle.

• • •

It was odd. I was a few hundreds of thousands years old, but I'd never been aware that I'd spent my life in darkness. The words *light* and *dark* had no meaning for me because I'd never been able to discern light before. Now that I saw it, I understood what I'd been missing that whole time. It was an unbelievable discovery. This changed everything.

My friends, too, were very enthusiastic.

"What a fantastic idea of the Creator!"

"Now we can finally see what he has made!"

"How did we ever manage without light?"

"This is indeed tremendous! Look, there, in the distance."

We'd already noticed how busy it was, but now we could see it too. Helium nuclei, protons, and electrons constantly shot by. It was as chaotic as a large city during rush hour, without traffic lights and traffic police. Imagine bicycles, scooters, and cars frantically crossing, shooting right past each other without taking anybody else into account, and multiply the speed by a thousand times. On top of this, picture the automobiles and buses not just coming from four but from every direction, including above and below—a three-dimensional traffic chaos with many collisions and near collisions. Unfortunately, being able to see what was happening did not mean that we could avoid collisions. Crashing, pushing, bumping, bouncing back and then speeding on, it was all part of what was happening. We fully participated in the process. It had become our daily routine.

This was the scene until something strange happened. We had yet another collision with an electron. However, instead of excusing herself and quickly going her own way, as would've been the norm, she stayed stuck to us—or, actually, she continued to circle around us as if attached by an invisible thread.

"Hello, up there, what is this?" I called out, irritated.

Before I could get an answer, there was a second collision with an electron. She, too, continued to stay in our proximity, annoying us like a mosquito we couldn't get rid of.

"Go away!" yelled Aris.

But they did not go. They felt particularly attracted to our friendship group. They circled around us like restless featherlight clouds. High above us we saw the one, then the other, floating past. They were like passersby, but they continuously traveled the same circular path as if they were on a three-dimensional merry-go-round, and we were the center.

"Isn't this fantastic!" one of the electrons exclaimed from afar. Her voice echoed as if we were in a gigantic, empty space. "We're a match made in heaven! Ha ha!"

"Well, I find it particularly bothersome," I said. "Can't we chase them away?"

"No," said Solon. "They can't detach themselves from us. The universe has cooled too much for that. But I don't know what we are going to do with them either."

I glanced around. "Now look," I said, "there are more atomic nuclei with the same problem."

"Just a minute," called Solon, "I should've known! This is actually very good news, everyone! We've had an upgrade. We are no longer a bare atomic nucleus! We've become a full-fledged atom. We needed this to become fully usable as a building block. I think things will start very soon now."

"The execution of the big plan, you mean?" I asked.

"What else? Prepare yourself for some spectacular changes, Pro!"

• • •

We soon got used to the presence of the electrons. They circled around us, but at such a distance that fortunately, they really didn't bother us, as they hardly interacted with us. Nevertheless, I didn't see very much of that so-called great plan.

To the contrary: the light that had made us so happy steadily became weaker, until it was hardly visible any longer, and finally it was extinguished. Once again, it became pitch-dark in the universe. That felt like a big loss, an amputation almost. There had been a time, of course, when I'd been content with the darkness. When I hadn't known any better. You don't miss what you don't know. But then came the light, a big surprise from the Creator, and now that it was gone, I missed it more than I could've imagined.

Was it something temporary, that light? Something that belonged to the beginning but that had now finished its task? Or might the Creator introduce it again?

I had no idea. *Was the Creator still occupied with his project*? I wondered. Actually, I couldn't find any evidence of it.

"Are you really sure?" I asked after a couple of million years. "About that plan?"

Solon, Aris, and Ensis knew it with certainty.

"Maybe you haven't understood it correctly," I suggested some ten million years later.

"No, we're certain of the plan."

"Nothing is happening. I think this is as far as it will go."

"You've got to have a bit of patience, Pro."

• • •

We were two hundred million years further down the line when the light came back! More majestic and brighter than ever. To be honest, I had not counted on it any longer. But I witnessed its return from close by, and it was the grandest, most special event that I had witnessed up to then. I know that the birth of a child always remains vivid in people's memory. And I will always remember the birth of our first star, the star that would become our home.

It was cold and dark at that time, so we didn't anticipate the event. What was the first thing we noticed? It became busier in our region, not unlike the rush hour at five-thirty in the evening near a commuter train station. After a little while, it became extremely busy. It was as if everyone in the whole city had the urge to go to the station and squeeze themselves into an already-full train, to set the Guinness World Record for "most people in a train car." This was just like what was happening in our place in the universe. The force of gravity pulled us and other atoms toward each other. While people get to a point where there is no room to crowd together any further, we kept pushing, with increasingly violent collisions as a result.

The cold was being chased away by an overwhelming heat. Something was happening. We lost our electrons in the chaos, and it seemed as though we landed in a melting pot with other protons and helium nuclei. It became hotter and hotter. When we reached the fifteen million degrees Celsius mark, the light came back in full force.

This wasn't flashes, not the pale light we'd experienced before (that in hindsight didn't amount to much), but an intense, blinding light that howled and devoured, burned and exploded, and shot out in all directions.

"What's happening?" I shouted over the roar of the flaming, elemental violence.

"Don't you see it?" Ensis called back. "Just look, those protons over there! They're melting together!"

Ensis was right. The nuclear force drove protons into each other's arms. They fused. Before our eyes, some protons changed into neutrons when they—with other protons—formed new helium nuclei. At the same time, an unbelievable, frightening, searing amount of energy was released in the form of blinding light.

"We're done for!" wailed Aris.

"Not at all," said Solon, and he gave Aris an encouraging nudge. "Watch carefully. They're changing, just as we did, long ago. See? Those over there look just like us. Hey! Hey everyone, are you all okay now? You've done very well!"

3

STARDUST

"It's so hot," Aris sighed.

"And that pressure," complained Ensis. "I can't stand it."

I'm not a complainer, but it was indeed unbearably hot, and it had been for a long time. Several hundred million years had passed.

The bright light of our star had changed into an orange-red gleam. Via the grapevine we heard of atoms in the outer shell—our star was becoming larger, expanding like a giant cake in a colossal oven.

"I'm wondering how long this can continue to go well," said Solon, who, through all this, had become a good friend.

"Perhaps this is the intent of the Creator," I called out above the noise of the violent explosions. "That we will fill the whole universe with our star."

"I don't know," Solon responded. "There are some strange things happening lately."

He was right. The heat and pressure had caused all kinds of new atomic elements to form, atoms we had never seen before—peculiar beings that considered themselves superior to us because they consisted of more protons and neutrons than we did. They also gave themselves grandiose names such as Neon, Vanadium, and Magnesium, which definitely annoyed me. But other things happened as well.

"Pro," said Solon when we had floated somewhat closer to the nucleus of the star. "I don't want to make you anxious, but something very strange is happening here. I don't trust it. Look over there, see what's coming our way."

"What?" I asked. I heard the ominous and serious tone in his voice. Then I saw what he was thinking about and understood that we were indeed in great danger.

We were obviously in motion, as everything and everyone was in constant motion. I know that humans have the idea that things can be immobile, but that's because they can't see clearly. If they could observe reality as it truly is, they would see that everything vibrates, shakes, rotates, wiggles, and jiggles, even things that from their viewpoint seem unmovable, such as trees, tables, and houses. Immobility does not exist. Immobility is an illusion.

At this moment, something caused us to begin to move at the same time as a helium nucleus that was next to us, as if an invisible hand saw to it that we were directed toward each other. This wasn't right. This was pure horror. To my dismay I noticed that we also continued to approach each other. There was no escape. The other helium nucleus came closer and closer and invaded our personal space.

"No!" I screamed. "No!"

"Watch out!" yelled Solon.

But it was already too late. In a flash we were combined into a new atomic nucleus, beryllium. And not the innocent kind of beryllium that you find on earth nowadays, in loudspeakers and x-ray equipment. No, this was an atomic nucleus that was short one neutron. Extremely dangerous! We were changed into a hand grenade whose pin had been pulled. Within a short time, we would explode. I prepared myself for the end.

At that moment something happened that I still consider to be a miracle. Yet another helium nucleus came to our aid. It came toward us with tremendous speed and reached us just in time. We clicked together as though we belonged to one unit, and in this way the explosive hand grenade had been made harmless.

"That was a close call," Solon said to one of the new protons. "Thanks, fellows!"

We were now in a large assembly. Personally, I found it a bit too busy, but Ensis and Aris were enthusiastic. They had the idea that we'd won the jackpot.

"Carbon! We've become carbon!"

"This is so cool."

"A promotion! We've received another promotion."

I looked to Solon. "Oh well," he said acceptingly. "Who knows what it'll be good for?"

"I'll bet the Creator has big plans for us," said Aris.

• • •

More and more helium bonded into carbon. Faster and faster it went, like popcorn in a popper: first one kernel popped, then another, and another, and then it accelerated quickly. Time and again I saw the same miracle occur: two helium nuclei came together into an extremely unstable beryllium nucleus that was stabilized at the last moment by a third helium nucleus, ensuring the whole assembly didn't fall apart.

In the meanwhile, we received disturbing messages from the outer shell of our star: increasingly more atoms were leaving our star and began circling it in large gaseous clouds.

"Traitors," Ensis grumbled.

"I can't blame them," Axion, one of the new protons, said. "I would leave too, if I had the chance. Everything here is about to explode. Our star seems to have reached the end of its life."

"What!" I cried. I was getting nervous about our situation. "How do you know?"

"I have my sources," said Axion.

"Such nonsense," said Ensis. "I don't believe any of it."

Axion snorted. "Don't believe it then."

To me too, it seemed to be a tall tale. I couldn't imagine life outside our star. Nevertheless, I was worried there might be a grain of truth in his story. And that proved to be the case. Our star was running out of fuel, and it was collapsing onto itself under the influence of gravity, which caused us to have some anxious moments. Imagine a sinkhole—a place where the ground gives way without warning, so that an entire house can sink away and disappear. But then much, much larger. A worldwide sinkhole in which everything and everyone disappears—trees, houses, apartment building, bridges, roads, entire cities . . . That's how it felt when our star collapsed onto itself. It seemed that the end had come.

Our star didn't give up easily, though. The heat and pressure caused it to revitalize, and with a heroic effort, it suddenly shook off as much material as it could, giving it a chance to carry on somewhere else in the universe. Then it contracted again. It managed to expand one more time, and my hope flared up again for just a short time. *Hold on*, I wanted to call out to it. *Hold on*! But what we saw were the final breaths of our dying star. After a while, nothing had remained of us but a small, white, glowing ball. A gigantically heavy ball, to be sure, but a small ball nevertheless. Even though this star—and in fact the entire universe—was as a feather in the hand of the Creator, a human being would need a forklift to pick up even one small spoonful of material from our star, as it weighed a few tons.

This could've been the end, but the Creator was not yet finished with our star. He likes recycling. By this time, I had seen enough of crashes and explosions. It wasn't as easy to throw me off kilter as before. Nevertheless, this was an extreme situation, even for me. Our dying star came back to life like a zombie. It blew itself up in a monstrous and violent burst of light and energy, more intense than I'd ever experienced and would ever see. The force of the explosion was unimaginably more forceful and destructive than the detonation of the most powerful atomic bomb. This was a blast of cosmic proportions.

Our star, our home, exploded. And we and all the other atoms in the star were thrown into space in an unbearable and blindingly bright light. When, on earth, an apartment building is blown up, the rubble crashes downward but the dust is blown hundreds of yards away. I was one of the dust particles that was blown from the building. I had no clue whatsoever where I would end up.

• • •

When I came to, it was dark and cold all around me, and I was totally disoriented.

"What's happening?" I asked weakly. "Where am I?"

"Hey, Pro! Is everything all right? You have to see this!" It was the voice of Solon. He was still there. All of them were still there, my carbon friends—six protons and six neutrons, and the six electrons that circled around us in a variety of orbits.

"What do you want me to see?" I questioned.

"Just look."

"I don't see anything. Where are we? Where's the light? Where's the star?"

"It no longer exists. It was blown to bits. Don't you remember that?"

A vague memory came back to me. The white light. The monstrous force that could not be withstood by anything or anyone. The frightening void and the darkness that we were thrown into. "But where are we then, Solon?"

"We are floating through space. Do you see it now? Do you see?"

I was thoroughly confused. "What? What are you talking about?"

"Those little points of light all around you. Just look."

It took a while before I understood what Solon meant, but he was right. If you took a good look, small light points could be seen in the dark, pinpricks, some brighter than others. They were all around us. The longer I looked, the more of them I saw. But it was nothing compared to the light that we'd been bathed in for half a billion years. "What are those little light points?"

Aris couldn't contain herself any longer. "They are stars, Pro!"

I laughed, thinking that she was making a little joke, but she was serious. She really thought that the small sparkling points of light in the sky were stars.

"Stars?" I questioned. "Do you really think that?"

"I'm sure," she replied, and Solon nodded. "They are stars. That's the truth."

This statement was so shocking that I could hardly grasp it. Our star, the center of the universe wasn't the only one? We weren't unique?

"Don't' you remember?" said Solon. "The Creator likes abundance."

"And that plan you mentioned?" I asked. "The plan the Creator was working on?"

"I think he's still at work on it," said Axion. "Wherever the Creator lets the light shine there are prospects for that plan to succeed."

"Too bad for us," I said with disappointment in my voice. "I had hoped so intensely that we would be used as part of the plan."

4

EARTH

Messy. Turbulent. Such was the place where we ended up, some eight billion years later. We raced with countless other atoms around a new star that had been born before our eyes, as it began to burn fiercely and put us in a blinding light. Helios we called him.

Collisions occurred frequently. One severe collision is etched in my memory as if it happened yesterday. That was the moment when we crashed into several atoms, clicked together like strong magnets, and could not possibly separate ourselves again. At first there was some bickering about who was at fault: "How could you possibly be so stupid!" "Move out of the way!" "Find someone else to smash into!" But then someone else announced, "Hey, we've had another upgrade!"

"What do you mean, upgrade?" I asked, annoyed.

"Just look," Solon replied. He sounded excited. "We formed a molecule! A molecule! And not just any molecule, no, we formed a *ribose* molecule." This meant that we had become a small village with more than two hundred residents. Although I have to say that one could hardly consider electrons as full-fledged citizens, because they were too small, too busy, and always occupied with each other, certainly now that they were circling around us in pairs. In general, the inhabitants of our village were able to get along well with each other, and there were always individuals to talk to, even though I got tired of them once in a while.

"Does anyone here know a little more?" inquired Lapis, one of the new neutrons. "Where are we going?"

"Nowhere," I said. "We're going around in circles, can't you see that?"

"Oooh, I see. I thought I recognized the area," Lapis admitted. "It seems like we're going faster and faster, though."

She was right. We spun and whirled, molecule and all, around Helios with increasing speed, and there was no opportunity to pull ourselves out of the circular dance. The star held us firmly in her grip, like a vinyl record that can't come loose from the turntable, no matter how fast it spins. Round and round we went, so rapidly that the cloud of stardust around us started to form a flat disk. More collisions followed, and we began to cluster with other molecules. After some time, we formed what you could call a dust particle. When we experienced another crash, we became stuck on a piece of stone that not much later, in turn, landed on a huge rock formation with a resounding blow.

At that moment, I had no idea that the Creator was completing his preparations and had begun building the house for the special creatures he had in mind. I had no idea that we would be allowed to be part of this, of the plan he had designed before the beginning of time: The solar system. The Earth. Life in an endless multitude of varieties and forms. The great adventure had begun.

• • •

It took another fifty million years for the solar system to acquire its final form. A large part of that time we whirled around the sun on our miniplanet, together with boulders, clumps of ice, moons, and other emerging planets. We sped around a racing circuit where no traffic rules were in effect, where no one took others into account, and where participants attempted to smash each other into oblivion. We crossed routes and collided with each other while planets and moons broke into pieces or had a change of direction. On our planet in the making, too, the damage was often significant. At times, the heat and pressure of a collision caused a rock formation to fuse with our planet, helping it to grow in size.

The violence came not only from the outside but also from within. Our constantly growing planet was no longer a massive clump of stone. It had become sort of like an egg—with a liquid, white-hot core where chemical reactions between minerals took place, and with a thin outer crust that showed cracks in various places. The earth often rumbled and shook. At times it tore open at weak places in the crust, spewing out molten stone. Gases that came to the surface from the inner core remained adrift around our planet and mingled with molecules that streamed in from space.

Due to numerous meteorites and comets, more and more ice was brought to our planet. The ice melted in the heat, of course, and slowly the surface of the earth became covered with water. It gradually became cooler too. Sometimes it was still unbearably hot, but there were also long periods of strong winds, hail, snow, rain, and thunderstorms.

Meanwhile, the daily bombardment of direct hits by meteorites and comets continued. At times, a meteorite hit would be so violent that the temperature on earth shot upward and entire seas evaporated. Then, millions of years later, the earth had a different appearance entirely—a sphere almost totally covered with water with an isolated island here or there.

We were fortunate to end up in the water near one of those land masses. I wasn't bored for a moment. So much was happening all around us! Besides, I never became tired of the view. To my big surprise, it became clear that the Creator had hidden many colors in the light. You could see them in a half circle in the sky when it rained with the sun shining at the same time. You saw them in the sunrise and the sunset, anew at every rotation of our planet. And at night, when we didn't see the sun, there was another surprise—a moon, which didn't shine itself but borrowed its light from the sun. Across the whole wide sky, we saw the other stars glow as small points of light, reminding me of things at the beginning.

I was so intrigued by everything around me that I only noticed when Aris pointed out to me, "It's become somewhat calmer recently, don't you think, Pro?"

I thought about it. "Now that you mention it, there are fewer direct hits than at first."

"Remarkably," said Ensis. "It's almost boring."

"Are the rock formations exhausted or what?"

"Ha ha!" an electron above us laughed. "You have no idea!"

"About what?" I responded.

"It's Jupiter!" she replied. "Haven't you heard?"

We knew nothing about it. Apparently Jupiter, a giant planet farther along in our solar system, with its enormous force of gravity, saw to it that all the rocks were prevented from crashing into the Earth as meteorites. Jupiter took all the heavy blows instead of us.

I long thought that all this was coincidence. I now see in it the care of the Creator, who in this way protected the Earth from the worst assaults from space, thus creating a calm place, a place for the home he wanted to make.

5

LIFE

"IMAGINE," ENSIS MUSED, "that you yourself could determine where you would like to go."

"How do you mean?" I asked.

"Imagine that you could think, 'I've had it with this ocean. I want to go on to dry land.' And that you could . . ." Ensis kept silent for a moment and then came up with a word we'd never used before, "travel, I would say."

"You have too much of an imagination," I chuckled.

The idea was too wild to be taken seriously. To be able to determine on your own where you would go? Nothing and nobody in the universe could determine where they would go. Protons, atoms, and molecules were all in constant motion; we all depended on external forces that moved us to new locations.

In the meantime, we had been underwater for several million years, and nothing much was happening. It seemed to me that we floated, weightlessly, back and forth in an ocean. It became dark and then light again, dark and light again. I became familiar with the sea, the currents, the tides, the alternation between cold and warm periods, the dark depths where no light could penetrate, but also the surface where you could see the starry sky, and sometimes, when we were lifted up by the waves, in the distance, the land.

"It seems as though more land is appearing," Ensis had already remarked earlier.

I had the same idea. For a long time there had been water everywhere, with an island here and there. But the continental plates that covered the glowing core of the earth constantly shifted, pushed against each other, and rose upward as if they were enormous ice floes. In this way more and more land was rising above the ocean's surface.

It seemed to me that our planet had changed significantly; but regrettably, we couldn't see much of that. But soon it would change. One morning we felt a tremor in the water, followed by an underground rumbling that announced yet another volcanic eruption.

"This is much too close!" I said, alarmed, when the water near us began to slosh around more wildly.

"Help!" called Ensis, in a bit of a panic.

"There's no problem," said Solon, reassuringly. "Just go with the flow."

Just at that moment, a stream of glowing magma shot upward through a crack in the earth's crust. The ocean swirled and churned, and we were shot out of the water in a cloud of steam, only to fall back again. But not into the sea—we landed in a pool of water, close to a beach.

"And Ensis," inquired Aris, "what do think of this? You were the one who wanted to travel."

Ensis kept silent, too impressed to comment. I was enthusiastic. It was as if we had landed in an entirely new world.

Because the pool was shallow, we had a good view of the surroundings. And since the water was evaporating, our pool soon became overcrowded. It seemed like a frantic dance party was in session, a party everyone was invited to—ribose molecules like us, but also many others. Hydrogen cyanide and a wild gang of amino acids, for example. Some of them looked rather simple; others were richly endowed with long tails or exotic clouds of electrons. We also met a strangely shaped lady who introduced herself as Ribonucleic Acid, “But you can also call me RNA.”

Because the dance floor was small, we constantly bumped into each other. Certain molecules got along with each other exceptionally well and began, after a kind of passionate tango, to clump together into fat globules with a sponge-like structure. But most of the dance partners by far went their own way again after a while, without much of a fuss. Some molecules, however, continued to party together and attracted other molecules and tried out various dance moves. We joined in too, of course. This was a party we couldn’t miss! Tirelessly we danced, millennium after millennium. Of course, there were accidents on the dance floor—not that this was a problem. A substantial collision could enable you to change partners and form a new molecular complex.

• • •

A long time later, some six hundred million years after the origin of our planet, I saw something remarkable happen. A group of clumped molecules began to change form spontaneously, as if an invisible person were sculpting them. The spongy sphere was first formed into a lemon shape, then into the shape of a dog bone, and after that into the shape of two snowballs stuck together. Then the most incredible event happened—the two spheres separated with a soft plop.

“Did you see that?” I called out, shocked. “Now there are two.”

“Four!” said Ensis.

“Eight! No, sixteen!” shouted Aris.

“Watch out!” warned Solon. “One is coming our way!”

At the next moment we were swallowed by one of those globules. I say globules, but from our perspective they were monstrous. It was rather like an ant that was sucked into an enormous zeppelin. We were pulled inside through a hole in the frayed exterior, and we landed in something that, more than anything, resembled a gigantic underwater labyrinth. It was a factory. Okay, factory is, in fact, much too big a word for the shabby, messy space where we found ourselves, but on closer scrutiny it did turn out to be a sort of factory. From the outside, building materials were brought in . . . or actually swallowed up. Once inside, they were chopped into pieces without mercy, or combined with other components to form new molecules. In this process, energy and waste products were released. The waste was removed and the energy was used.

To my surprise, RNA, the strangely shaped lady I had met earlier at the dance party, turned out to be the manager. It was clear she was firmly in control. She saw to it that the factory was operational around the clock, and she directed all the chemical processes.

Thanks to her special personality, she did something else. At first sight she did not look very attractive; she looked like a crumpled-up hand-knit scarf. When she stretched out, however, you could see what she really looked like—a long cord with thousands of molecular letters strung together, letters that formed a code. That code was the instruction manual that kept the factory running. Our manager saw to it that new RNA molecules were formed from atoms supplied from the outside. In this she copied her own code into those molecules: they were little computers that contained the information that they needed.

"Watch out!" someone called. "We're going to be split!"

The factory began to stretch itself out, changed in shape, and in the next moment a part of the factory separated itself, split away from us, and continued as an independent entity.

"What is all this?" I asked in surprise. "Where did I end up?"

"This is a 'protocell,'" a molecule told me.

"And the piece that we've lost?"

"That's a daughter cell. It'll continue by itself."

"As . . . as what?"

"Also as a protocell. What did you think? Excuse me, we have to get back to work." From behind a small wall a kind of a chopping machine came toward us.

"No!" I yelled, but it was already too late. Our beautiful ribose molecule was chopped into two pieces. A garbage truck came to get the piece that was apparently considered waste by the manager. I can still hear the frightened voices of my friends: "No, No! Not us!" We who were left were carried off to another place.

"What's happening?" Ensis asked anxiously. "What are you planning to do?"

"You're going to move to a higher position," I heard someone say. The next moment we were clicked as a train car to some other train units, and code letters were attached to the sides. Altogether this only took a fraction of a second, but it seemed much longer. We had become part of a new RNA molecule.

What had just happened was in no way comparable to the severity of things we had experienced earlier. The explosions and eruptions, the nuclear fusions, the collapse of our star, the violence with which our planet was formed, the meteorites, volcanoes, and seaquakes . . . until now it had been quite an adventurous journey. Nevertheless, I knew that something really special was now taking place; that we were entering a whole new and unknown world.

This was no longer a deadly boring place. This wasn't the lifeless movement of isolated molecules. No, here I saw the beginning of life. Life that collected energy and raw materials from the exterior; that passed on life to offspring; that gathered information and passed on knowledge. Here a living being arose. It escaped the natural tendency to decay. This new being struggled against the chaos that had the universe in its grip. The first cell.

6

EVOLUTION

IT WAS A WHILE LATER, some five hundred million years, I'd guess. We no longer lived in a small, simple factory but in a busy metropolis, a city where the inhabitants were not to be trusted. Our world was now a pandemonium, like the mosh pit at Woodstock '99 with a wild and delirious crowd. We were pushed, pinched, and pulled away to all sides. Molecules constantly tried to lure us along, grabbed us, and tried to rob us. Some molecules blatantly chained themselves to us and joined us on our way, until another came along and dragged us away because we were needed elsewhere. Obstacles were everywhere; threads and passages ran all over the place, and there wasn't a moment of rest to be had. Apparently this was all necessary in order for the factory to function efficiently. Over the years, the descendants of the first cell had become increasingly more skillful. They managed to produce a sort of outer covering, a soap bubble-like layer that provided protection but also allowed needed substances to get through. The cells had developed a method of obtaining energy from the outside to keep the factory operating by using energy from the sun. And all that information was always copied to pass along to their offspring.

This copying didn't always go perfectly. Imagine a copying apparatus that's been set down in an awkward place—in the middle of a busy passageway—and receives a hefty blow every once in a while. With all these blows inside the cell, the hereditary material would also occasionally get some significant hits. But that proved to be exactly what the Creator had in mind. The mistakes in the copied material provided variation and development. The cells began to change.

Most cells hardly noticed the copying mistakes, while in others the changes proved to be fatal. But one or two benefited from the tangle of changes in their particular environment—a better operating system, a more convenient way of taking in nourishment or carrying off waste. And those changes would then be passed on to offspring. Some unicellular beings were formed that were so advanced they could venture out on their own and swim across the oceans.

• • •

My friends and I roamed for a long time through the seas in many different single-celled vessels. We saw how the underwater world became populated with these beings. They were everywhere and they learned more and more. One of the discoveries they made was that you don't necessarily have to copy yourself exactly to have offspring. You don't have to make clones of yourself. You could also exchange information with an individual of your kind so that you would have offspring with other features. That was a huge breakthrough. The management also changed to a new owner, an extremely long, slim molecule that called itself DNA and took over the control of the hereditary information. Over time, more and more details were added. And finally, some three and a half billion years after the origin of the first cell, the first multicellular beings arose.

I was there. I saw it happen. I was even part of it. It was a miracle of beauty and ingenuity—cells began to cooperate, joined together, began to form *one* organism and, with good mutual agreement with neighboring cells, decided whether they would form into a stalk or antenna, spinal cord, or wing. If it was needed, they were even prepared to offer up their lives.

Worm-like animals, algae, fungi, and even fish came into being. It became more and more beautiful underwater. At times I thought back to when the star was our home. But I also remembered the terrible and frightening moment when the star exploded, and we had been flung into space. I had now found a new home, and it was better than I could ever have imagined.

"I've heard something so ridiculous," Ensis said one day.

"What did you hear," I inquired.

"I've heard that life out of the water is possible."

Solon and Aris couldn't stop laughing, and I must say I found the idea rather amusing myself. Life had taken its course in the water for more than three billion years. Everything was adapted to it. This is what the Creator must have had in mind—this rich, beautiful, underwater world with troughs and sandbanks, corals and kelps, jellyfish and octopuses, and fish in all kinds of shapes and colors.

"How could something or someone live out of the water?" I asked.

"Right, how would a creature see to it that it didn't just drain away?" grinned Solon. "Just take a jellyfish. When it lands on a rock out of the water, it dies. It dries out in no time!"

"Besides, there's nothing to eat on land," Aris added.

"How would an animal move, out of the water?" I asked. "On land it wouldn't get anywhere using its tail. And how would it reproduce?"

"I know. I can't answer your questions," said Ensis, "but that's what they're saying."

In principle, of course, there were enough building materials on land—the same atoms and molecules as in the ocean. We'd lived there ourselves for a while.

But for living beings it was dangerous out of the water. It was just as impossible to live there as it would be now for a human being to live on Jupiter. Without a space suit. Without protection.

Yet, Ensis was right. There were plants that had discovered a smart way to pump nutrients up from the earth via a root system. And they managed to take in CO_2 and secrete waste products through small openings in their leaves. There were some animals that had developed something similar: a pump to help nutrients and liquid flow through their body, small tubes though which the liquids could reach all parts of their structure, hard components that could support their body parts. They could also move out of the water and not be helplessly lost like squids or jellyfish that washed up on the beach. They had a swiveling neck, a firm skin serving as protection and keeping the liquids inside, and a method of bringing in oxygen.

The Creator had seen to it that, in addition to the ocean, the land had become habitable.

• • •

One day we went on land by means of one of those exceptional animals. We were part of a fish that could breathe under water as well as out of the water: under water it used its gills, and on land it breathed using its lungs that were protected by an armor of ribs. It had fins that could function as legs with which it would, at times, pull itself onto the shore.

I thought it was tremendous, not only because I could see the sun without the water's filtering it, but also because the surface of the land was completely different from what I remembered. The land no longer consisted of bare rocks. Now it was covered with green, with ferns, grasses, and dense forests.

There was so much to see that I was sorry that our "landfish" always chose to eventually return to the water.

A couple of million years after the landfish had disappeared from the scene, we ended up in a sort of lizard that lived on land. When it died and decayed, we became part of a fern. Subsequently we were taken up into a shoot that slowly became larger, grew above the ground, and eventually becoming a giant clubmoss tree. From the inside I saw how the tree sucked nutrients from the ground and managed to transport them to the highest leaves in the crown. Not only did I see this, my friends and I actively cooperated in the process as if we had never done anything else.

When looking at the law of gravity, this would seem impossible—water wants to go down, back to the earth. But after millions of years of failures, of trying again and again, plants had learned to do it using an ingenious system of varying pressures, using tubes that were so narrow that the water was not bothered by gravity.

• • •

We lived many lives, worked in numerous "cell factories" and I became more and more impressed by the exuberant creativity of the Creator. His signature was to be seen everywhere: In the beauty of DNA molecules that passed on their information from cell to cell until they finally ended up in the soil after the death of an animal or plant, where they then served as a nutrient source for a living creature to follow. In the tasks he gave each living being. In the countless possibilities he had incorporated into his creation. Every atom, every molecule, every cell, every plant, and every animal showed something of his exorbitant creativity: the amoebas, the jellyfish, the squids, the gigantic dragonflies that skimmed over the water, the armored scorpions. But also the ferns, the flowers, the trees, the grass with which he had covered the earth, the seas, the rivers, the snow that covered the mountains. The multicolored diversity was overwhelming. What pleasure the Creator must have had in all of this. He'd made something unbelievable out of this small rock formation in his gigantic universe, a planet that bristled and sizzled with life, with shape and color, with beauty and ingenuity, a place where all things were in tune with each other.

• • •

"I heard that he has even more plans," Lapis announced one day.

"What kind of plans?" I asked.

"I don't know exactly," said Lapis. "It seems it has to do with the great plan. The Creator wants to make personal contact with one of the beings he has made."

"Contact?" By now I was getting used to being surprised time and again, but this was a turn I hadn't seen coming.

"With a being that's going to be like him. And he wants to enter into a special bond with that being. It sounds like he's been looking forward to this for billions of years."

A being that would resemble the Creator? I couldn't imagine anything like that. Were we dealing with a terrestrial animal or an aquatic one?

Lapis didn't know either, so from that moment on, we watched all creatures closely and we discussed what it could be, the being that the Creator had in mind. The landfish maybe? No, that couldn't be. We hadn't seen them for a long time. I didn't even know whether they still existed. The octopus then? The lizard?

"This must be it," Ensis said one day. The animal we were now part of was bigger than all creatures on earth. It was almost thirty-six feet long, eighteen feet tall, and walked on two legs. It had powerful claws and enormous teeth, and it devoured everything that crossed its path. "This is certainly the being that the Creator wants to be in contact with."

"I rather think that *this* is the special creature, Ensis," I said a couple of million years later. We ascended from the earth and looked down on the coniferous trees below us and on the water that glittered in the sun. The animal that was our current home certainly wasn't the largest of all animals, but it had a fascinating ability—it could resist the force of gravity. At the place where other land animals had front legs, these animals had large appendages they could use to float on the wind; the front legs of its ancestors had been gradually changed into wings.

"This is indeed spectacular," Ensis admitted, "but haven't we seen animals that could fly before?"

"Those were insects," I said. "This animal is much larger and more complex. Can't you see that?"

"I don't know, there are so many animals, and they all have something unique. Each one could be it. What do you think of that animal over there?"

I looked at the small mouse-like rodent that was rummaging between the ferns. I'd seen it before, but it hardly seemed worth looking at. So small. So weak compared to many other animals. It had no armor, no scales, or shell . . . only a hairy coat.

"It looks strange on the outside," I said. "To me it looks rather vulnerable."

"It can keep itself warm!" Ensis said.

"I don't see what's so special about that," Solon said. "Some reptiles can do that too."

Ensis was not to be discouraged. "The female of this animal can feed her young with a liquid from her own body. And this little animal is clever, cleverer than all the other animals that have existed until now."

In the next instant our host struck. It let itself fall from the sky, seized the little animal with its claws, ground it between its teeth, and swallowed it.

7

CATASTROPHE

IT IS POSSIBLE THAT I've given the impression that things were peaceful on earth in that time. That was not the case. It had certainly become much quieter since Jupiter had become our guard. The daily powerful bombardments from space were a thing of the past. Of course, there were still lightning strikes resulting in enormous forest fires and extremely hot and extremely cold periods. Occasionally meteorites escaped Jupiter's security, meteorites so large that, at times, life around the crash site was destroyed. But that was minor compared to the Great Disaster that occurred some four and a half billion years after the earth originated.

I didn't see it coming. No one saw it coming. A block of stone with a six-mile diameter came down from space. At forty thousand miles per hour, it crashed into the atmosphere and struck a peninsula with the power of billions of atom bombs, near what would later be called Mexico. This catastrophe radically changed the course of the history of the earth and its inhabitants.

All life around the crater was wiped out in an instant. At the place of impact, the pressure and heat melted the stone to liquid glass. Drops of glass were blown upward into the upper reaches of the atmosphere, where they cooled and solidified, falling back to earth in a downpour of glass pebbles. Fish in rivers two thousand miles away unsuspectingly took in the particles and their gills became clogged. If this wasn't the cause of their death, they died soon afterward in the tsunami that followed the impact and covered everything with a layer of mud that was dozens of feet deep. From Mexico, the wave, which initially had a height of one mile, swept over the land, destroying everything it encountered on its way.

"This is the end," Axion shouted, "the end of the world!"

This sounded somewhat dramatic, perhaps, but it expressed precisely what I feared as well. It indeed seemed that the earth would perish. Red-hot stones and dust darkened the sky. Burning sulfur rained over a wide area and burned all animals alive. The forest fires that were ignited everywhere polluted the air with soot and ashes, so that sunlight could no longer reach the earth. It was pitch-black for two years, and it became horribly cold. When the light slowly returned, the full extent of the disaster finally became clear.

The great dinosaurs were gone. Those that didn't perish at the impact, the tsunami, or the downpour of glass and sulfur died because they couldn't find any food. The darkness killed many plants and trees; there was no sunlight. The results were mind-boggling. The herbivores starved and perished. As a result, the carnivores were doomed too; there wasn't enough prey for them to eat. Three-quarters of the mammals became extinct. Of the dinosaurs, only the birds escaped the carnage. It was a terrible slaughter. All that was delightful on earth seemed to have disappeared.

"All those beautiful animals," Ensis mourned. "They are no more. The way I see it, the Creator's project has failed."

"How can you say that?" Solon said indignantly. "The Creator will not let his project fail."

"Don't you remember how beautiful it was before the disaster?"

"I know how beautiful it was. I admit it indeed looks rather pitiful compared to what it used to be. Still, I can't imagine that this will get out of hand. Perhaps the Creator needed this step to make something new, something even more beautiful. There is a time for everything, I've heard. A time for dark, a time for light. A time for cold, a time for heat. A time for large animals and a time for the small ones. Look! Look at that small animal there by that fern."

It was a rodent. It looked like the small animal we saw many years ago. The animal was pathetically small, nothing compared with the 110-foot Supersaurus, or the Giraffatitan that could look out over the trees. This rodent had survived the Great Disaster. It escaped by the skin of its teeth. Because it was so clever? Because it was able to hibernate? By pure luck? Whatever the case, it was still alive and nibbled on a plant shoot emerging from the earth.

• • •

We regularly changed our home—at one time we viewed the earth from the ground, at another from a bird, a plant, a rodent. In an endless cycle we were taken up by plants, eaten by animals and eliminated, burned in a forest fire, blown away by the wind.

I looked with amazement at how the dead earth began to produce new life. How the green wrestled itself from the soil, twisted itself upward along rock faces, toward the light—grasses, berry bushes, palms, leafy trees, seed-bearing plants. I saw how the sky was filled with all kinds of birds, descendants of the small flying dinosaurs that had survived the disaster. How rapidly—within half a million years—numerous new species of mammals came into being: great and small, sluggish and swift, weak and strong. The small animals often served as food for the larger ones, which in turn were prey for even stronger and larger animals. Now that the dinosaurs had died out, other animals finally had a chance. We saw them graze and gnaw, but also hunt, fight, and kill. Sometimes we saw a feline eat her own young or a mother bird push the smallest nestling out of the nest.

"How can she do that?" asked Ensis, who wasn't always the most clever in his comments.

"There's no other way," I said. "Don't you get it? Life has to be passed on. If there isn't enough food for all the young, a mother has to choose those that have the best chance to grow up. If she doesn't, none of the young has a chance. They all die."

"Exactly," said Solon. "Life has to be passed on. That's what matters."

Distant relatives of the small rodent developed handy toes and fingers they could use to climb trees, and eyes that could judge distances, which was useful for jumping. In fact, a number of other kinds of animals developed this type of unbelievably precise "3D eyes." It was as if an invisible hand pushed them all in the same direction.

"Those tree climbers," I said one day. At that particular moment we lived in a fruit tree and saw, close by, a mother animal with a little one on her back. "They sure have a large head compared to other animals. That seems so inconvenient to me."

"Large brains," Ensis replied.

"Well, I can understand that," I said. We had been in a skull often enough to know what that looked like inside and how it functioned. "I know what's inside a head, but brains require a lot of energy. I wonder what having such a large brain is good for. It seems to me that you are better off if you have more muscles."

I considered the dinosaurs, many with such small heads. *An animal doesn't need a large brain*, I thought. *This tree animal searches for food a good part of the day just to keep its brain functioning.*

I'd noticed it often—the copying mistakes that caused changes in the DNA were not always useful. At times they were even a deterioration, when it made the animal slower, for example. Deteriorations could even cause the extinction of an animal species. That had happened numerous times.

But the tree climber did not go extinct. This animal developed further and further and had descendants that were larger and proved to be even more clever. The primates.

They lived in groups, protected each other, helped each other, fought with each other. They learned whom to trust and whom to watch out for. They discovered that it can be rewarding to do something nice for someone, but also that sometimes you have to stick up for yourself.

Some of the primates developed into gorillas, some others into chimpanzees. Yet another branch developed into hominins of various kinds that became more and more clever, walked on two feet, and began to make stone tools. They came, they lived, and became extinct. All of them except *Homo sapiens*, human beings.

Would this be the one the Creator had in mind? I wondered. But I had asked myself that question all too often, so I kept watching the other animals carefully too. Bees communicated through different types of dances to show other bees the direction and distance to a field of flowers or a water source. Lions, thanks to a combination of power, suppleness, and speed, were at the top of the food chain, earning them pride of place in all bestiaries; no animal was stronger, no animal was a threat to them. Or elephants, that formed such a tight-knit family group, that transferred knowledge and experience to younger animals, that played, used tools, and mourned for long periods when a family member died. They seemed to have self-consciousness and empathy, could solve complicated problems, and communicated all kinds of information to each other in their own language.

Still, the more I saw of *Homo sapiens*, the more I got the feeling that this might be the creature that the Creator had long had in mind. Through fortunate circumstances, we became part of the world of *Homo sapiens* more and more often—a butterfly, a mold, an ostrich bone that he used as a knife—and so, at times, we witnessed the developments close-up. Human beings discovered the secret of fire, they learned to work with iron and copper, they began to make paint and began to tell their stories by making detailed paintings on cave walls; they made musical instruments and built huts. They also learned to make weapons for the hunt, with which they also regularly attacked each other. Besides all this, they developed an increasingly complex language that enabled them to collect and pass on more and more knowledge. They could explain to each other in great detail where to find food, how to hunt, which plants were edible. And they began to tell each other stories. About the sun and the moon, about trees, mountains, and seas. They told each other about the past, about how the world began, about spirits and gods.

"They are searching," I said one evening after I heard an old man tell such a story as we sat around the fire.

"Yes," replied Ensis. "They are searching, but they are none too clever. They see their fellow creatures—the sun, the moon, the river—as gods. They still have no idea."

8

EDEN

I SAW HIM COMING TOWARD ME. A well-muscled young man with dark skin and a black beard. On his upper arm he had a band of braided grass. He walked straight over to a woman at work in the garden. She was young and slender, her breasts were those of a woman who had never nursed a child, and around her neck she wore a necklace with colored beads. Her face lit up when she caught sight of the man.

"Womuntu!" she cried. "You are back. I thought you would be gone for three days."

Together with my friends, I had become part of a bamboo cane. We were a small segment of the fencing around the vegetable garden. The man stepped across our bamboo cane into the garden. "I've come back early. Something has happened. I have to tell you about it, Maisha."

"I have to tell you something too, something you just won't believe."

"Do you know who I have met?"

"Early this morning I . . ."

They both talked at the same time, so they didn't really listen to each other.

"Okay," said the man. He laughed. "You go first."

"No, you first," said the woman. "Tell me!"

The man sat down on a rock and gazed at the mountains in the distance and the clouds that floated by in the blue sky.

"Something happened," he said. "Something very . . . something that . . ." He stopped, put his hands in front of his eyes, moved them away again, and took a deep breath. "So yesterday," he continued, "I'd floated a long way downstream. At sunset I'd just made it to the big bend in the river. I moored my canoe, made camp there, and built a fire to keep the wild animals away, those big preying cats with manes. After that, I walked to the canoe to get the fish I'd caught on my way. All of a sudden there was someone beside me. A man. A stranger I'd never met. Where he came from, I do not know. I know for sure that I was all by myself before I noticed him. But there he was. 'Good evening, Womuntu, son of the earth,' he said. It gave me such a fright that I became weak in the knees and collapsed. How did he know my name? I was overwhelmed. I don't know where it came from, but I had this deep feeling that this was not just anybody. I sensed he was greater than I was, older than the oldest of our ancestors, stronger than the strongest warrior, and wiser and more righteous than a human being had ever been. I thought my fate had been decided. But he laid his hand on my shoulder and put me at ease. 'It is me,' he said. 'I who always was.'"

The man paused for a short time and then continued. "I can't believe it myself, but I think it was the Creator of heaven and earth himself, Maisha. I met the Creator."

"Did you hear that?" I exclaimed excitedly. I couldn't comprehend it. This had never occurred in the nearly fourteen billion years from the time we had come into being, from that small seed, that minuscule egg that contained everything: time and space, all energy, all building blocks that would be needed (including my friends and myself), the laws, light and dark, heat and cold, life, and all possibilities for an infinitely large universe.

The Creator who stepped into his own creation! This was even more mind-blowing than the origin of light, greater than the birth of the first star, and more stunning than the origin of life. I had never given it a thought, how the Creator could make contact with the beings that he had in mind.

"He came to them!" cried Ensis, as excited as I had been.

"Shush," said Solon.

I said nothing and listened in amazement as Womuntu spoke about what he'd experienced. How the Creator had taken him on a walk along the river. "He walked alongside me like a father," he continued, "as a good friend. And we talked together. For hours. It was as if I had always known him. It was like a dream, but it was not a dream. It was really him, Maisha, the Creator. The one we have been searching for the whole time. The one we talked about when we sat around the fire. The one we thanked when it rained. Whose presence we felt without knowing him. The one we saw in the sun, the moon, and the stars, in every baby that was born, in the harvest, and in all new life."

The man shook his head as if he still couldn't believe it.

"So I stepped into my canoe, early this morning, and I rowed back, as fast as I could. I had to see you, Maisha. I had to tell you."

"Let me tell you something," said the woman. She smiled, "I met him too."

"What?" said the man, who appeared to be as surprised as I was. "Really? Where? And when?"

"Here in the garden, early this morning, when the sun was still behind the hill. I was awake early and couldn't sleep, so I came over here, as I usually do. You know how I like to be in the garden early in the morning when there are no other people, when it's still nice and cool. The moment when the first birds wake up and begin to sing, when the day comes alive.

"And then he was there. I didn't see where he came from either. Suddenly he stood beside me. I wasn't startled, I don't think. It may sound ridiculous, but I had the idea that I'd known him for a long time. And that he knew me. The sun rose above the hill, and I pointed him to all those colors: orange, pink, purple, dark blue . . . I told him how beautiful I thought the sunrise was, and he laughed and said, 'Thank you.' I looked at him. It took a little while, but then I realized what he'd said and who he was. Who was standing beside me. Then I felt frightened too, I think. I had all kinds of feelings—fear, confusion, reverence. But happiness especially, a deep happiness I'd never experienced before. 'My Lord,' I cried out, and I bowed before him. 'Is that you?'"

"So you too! You talked with him too? Unbelievable!"

"He made the sun. Did you know that, Womuntu? He conceived everything—the light, the colors, the animals, our ancestors, us! He told me that he'd been longing for this moment for such a long time. That he'd thought of us even before he made the light and the colors."

"Yes!" called the man. "That's what he told me too! That he loved us already before he made the stars. That he'd been so passionate when we came into being that it was as if he himself had literally sculpted us from the earth and blown the breath of life into us. And that he had purposely made two equal versions of us, male and female, so that we wouldn't be alone and could complement each other and reflect his love together."

"I have the feeling that I slept all that time, and that I've only just awoken."

"I feel the same!" said the man. "Everything looks different today."

The woman got down on her knees in the grass. She stuck out her finger and let a small, shiny blue beetle crawl from a blade of grass onto her finger. "I now see it in everything—in the grass, the insects, the trees, the flowers and butterflies, the water, the sun. They all call out the greatness of the Creator, the beloved, the source of our life. Why didn't we hear this before?" The beetle reached the top of her finger, spread its wings, and flew away.

"We knew that there had to be something or someone behind it all," said the man, "but we didn't know who he was."

The woman jumped up. "That we thought it was the sun! And that the moon was also a god. How could we have been so blind?"

"He isn't upset, and he isn't angry that we were so unaware. He loves us, Maisha."

"We saw him in the greatest things we could see or imagine, but he proved to be much, much greater. The universe is like a speck of dust in his hand."

• • •

"Do you hear that?" I asked. "A speck of dust, a particle! The universe is like a proton!"

"Or a neutron," answered Ensis.

"Be quiet!" said Solon. "I want to hear what they're saying."

"All the water in the rivers and the big sea . . ." said the woman, "it's no more than a drop for him. How is it possible that he came to us, to you and to me?"

"I don't know. There's so much I don't understand. But one thing I do know, he has a task for us."

The woman nodded enthusiastically. "That we look after his earth! As caretakers."

"He said that to me too! We're the only creatures to whom he wanted to entrust this, he said."

"But this is an enormous task, Womuntu. The animals, the plants, the water and the land, the forests and the sea . . . that we have to care for it all. I think it's unbelievable that he entrusts us with so much."

The man took a deep breath. "I still can't comprehend it all. I still feel so young, so inexperienced. This is too great a task for us."

The woman took his hand. "He wouldn't ask us to do something we can't do. We don't have to do it all ourselves, do we? This is a task for all of us humans together. He'll help us with it. We can always ask him for guidance. He said that he would visit regularly, remember? If anyone knows what needs to be done, it'd be him."

"You're right. Come on, Maisha, let's go to the village. We have to tell the others. This will change everything!"

"Yes! And we have to tell them about the two wells. Did the Creator mention that to you?"

"That there is one well that gives life and one we should not draw water from, you mean? Yes! He did tell me. We've got to get everyone together and explain to them what it means, so no one will make any mistakes."

The man helped the woman over the fence, stepped over it himself, and picked a mango from the tree near the fence. He took his bone knife to peel it and gave the woman a piece of the fruit. While she took a bite, he pulled her close and kissed her. "I'm so glad to see you again," I heard him say. "I'm so glad you're home again," she answered with a smile. They walked hand-in-hand toward the village. "Lion," I heard the man say before they both disappeared beneath the trees. He nodded toward the animal that was asleep under a shady bush in the distance, its majestic head resting on its paws. "Wouldn't that be a good name for the large predator cat with the long mane?"

So *this* was the being that the Creator had in mind when he took the barren clump of rock that had begun as a piece of waste from a star, and let it grow out into the unique, incredibly beautiful planet of which I was a part. This descendant of the first cell. And of the fish. Of the rodents. Of the apes. I was deeply impressed.

• • •

"What I do wonder about . . ." Aris said later, when the night had fallen and we were looking at the stars, our former home, now so far away and unreachable. "Why would the Creator have chosen this particular being? If you look at its genes, you notice how very little they differ from those of other animals. What then makes this creature so very different?"

"This creature can think about itself," I said. "That's what I've noticed. That this being is really self-aware."

"There are other animals that seem to be aware of themselves," said Aris. "Dolphins. Crows. The great apes."

"If that's the only thing, could it just as well have been a crow?" Ensis asked. "I don't believe that. This being truly *does* rise above the others. Look at the way they communicate. It's a lot more complex than in any other animal species."

He had a point there. In some way or other, humans were able to express thoughts and feelings in sounds with thousands of variations, in words. This, I realized, made it possible to communicate with other humans, to make plans, and to reason with each other. I'd often seen how humans passed information to each other about which plants were edible, the best ways to hunt, and how to make a fire. Together they were gathering more and more knowledge; they were always becoming increasingly clever.

"What seems to be even more important," said Solon, "is that this is the only creature that thinks about where it came from, about the origin and meaning of life. These human beings don't only live in the moment, they also think about the past and the future. They love. They paint. They make music. Personally, I think this may well be the creature that can make or break the planet. I don't find it strange at all that the Creator gave them this special task."

"For my part, what I find really striking," said Ensis, "is that these beings, people, don't always follow their instincts. They can control themselves. At times you see them get angry, but they don't attack. In this they clearly differ from other animals."

At that moment I thought of something; something that might perhaps be the most important. "As far as I can tell, a human being is the only creature that looks skyward and realizes that there must be a Creator, someone greater than themselves."

"You're right," said Solon. "That is indeed remarkable, Pro. This animal species has understood that there must be a power who made the stars, the sun, and moon, who makes the rainbow appear and enables the crops to grow."

"Even if they had no idea who he was until this moment," Aris said with a bit of a sneer.

"No," I said. "And I ask myself whether they would've ever figured that out if the Creator hadn't come to earth to meet them in person to tell them that it's *him* they'd been worshiping, without knowing it, to invite them to begin a new community, and to care for his creation as his stewards."

For a moment I was quiet, and then continued. "If you ask me, this changes everything."

"You're right, Pro," said Aris. "This is unique. I think that we have seen a new species emerge today. *Homo sapiens* has become *Homo divinus*."

"Sorry, what?" I asked.

"*Homo divinus*. Human beings that have been addressed by God. This makes them forever another type of being, a unique species."

9

WELL

THE NEXT MORNING WE SAW THEM AGAIN, Womuntu and Maisha. This time they weren't alone. They had taken their whole group along—men, some with a toddler on their shoulder, others carrying a fishnet made of tall grass. There were women with babies in their arms, old people, and children that ran around the group, playing tag with each other, and then ran away until they were told to behave themselves by their mothers. There was a feverish excitement in the air and busy talk and laughter.

When Womuntu climbed onto a fallen tree trunk and raised his hand, it gradually became quiet. "Dear fathers and mothers," he said, and his face beamed. "Brothers and sisters. Yesterday evening Maisha and I told you about a special meeting that we had. We told you about the Creator that we've sought for so long but had never seen. That we groped for without being able to touch.

"Finally we know who it is that made darkness and light, who stretched out the firmament, who separated land and sea, planned the birds, the land animals, and people. We finally know who hung the lights in the sky, makes the rain fall, lets the plants and trees grow, and gives us food to eat."

A man raised his hand. "I thought about it last night, Womuntu. I still don't get it. From our ancestors we learned that the sun is our god and that the moon is a goddess."

"That is the good news!" declared Maisha. "The sun and the moon are just two lights that the Creator has made."

"But what about the river then?" a woman asked.

"All waters are made by the Creator," Womuntu answered. "They come from him. They flow because he lets them flow. The river bursts through its banks and irrigates the land when the Creator calls. All the things we thought were gods, they are creatures, just as we are. The sun and moon go in the orbits that the Creator has appointed for them. The rivers clap their hands for him, the mountains call out his greatness. Everything we see, he has conceived it, he has made it, he upholds it. And that means that we can live without fear and without worries." Womuntu pointed to a couple of toddlers that were playing with sticks and pebbles in the grass. "Just like these children here. The Creator loves us, as a mother loves her child. He protects us as a father protects his family. Everything that all of you can see, he gives us, he lends to us and allows us to use it. He wants us to care for it on his behalf. We may enjoy it, and we may use everything that we need. We may sail on the seas, we may hike in every forest and climb every mountain, we may draw water from every well—also from the well that is the source of our river, upstream in the east."

"The sacred well?" an old man said, with surprise in his voice.

Womuntu nodded, and Maisha now began to speak. "Yes, also from the sacred well. The Creator has given special powers to this well. When we draw from it and drink the water, we will live with him forever."

"We'll go there!" a boy cried out. "Let's go!"

"I'm coming too," his sister said.

Womuntu laughed. "Wait a moment," he said. "I told you yesterday that there's also one well we must avoid, from which we may not draw. That's the well at the base of the hill east of the village. The Creator has said, 'Drink from all the wells, streams and rivers, but do not drink from the well at the base of the hill. When you do that, you break the bond of love and you shall die.' That's why we're here. I will show you that well today so that later you can also tell your children and grandchildren. Fathers and mothers, brothers and sisters, let's consider that well to be a prohibited well, the well from which we may not drink. The Creator was very serious about this. It is the one thing we're not allowed to do."

I saw the people nod seriously.

"Let's put stones around it," one man said. "And let's put a mark on these stones so that we know this is a well we must avoid."

"Or we'll shut it down. Even better."

"I don't even want to get close. At the foot of the hill, you said?"

A boy around ten years old took a step forward. He had an expression on his face that I'd come to recognize. He was scared.

"Ardhi," said Womuntu. "Speak."

The boy hesitated for a moment, then said softly, "Womuntu, a few weeks ago I was at the foot of the hill. I'd walked a long way and I was thirsty." The boy looked down. "I drank from the forbidden well. Am I . . . am I going to die?"

Womuntu smiled at him. "No, my son, don't worry. You didn't know. Besides, the Creator hadn't yet said anything about it when you drank. Where there is no law, there is no violation."

"The Creator doesn't want you to be afraid," added Maisha. "He doesn't intend to bring us trouble. He wants us to choose the right way and for us to be happy. That's why he warns us and why we want to show you the well, so that we can imprint all of it in our memory. After that we'll leave the place behind us and go back home."

"Good idea, dear," said Womuntu. "Let's visit the well, and after that we can celebrate. This is a day for a big party, brothers and sisters, fathers and mothers. This is a day to make music, to sing and dance for the Creator of heaven and earth. To thank him because he has let us know who he is. Because he loves us. Because he has such beautiful plans for this world, for us."

"Let's go," said a man, "then we're done."

"First let's put our fishnets out in the arm of the river," said Womuntu. "Then we'll know for sure that we'll have fish when we return. Ardhi, will you help the men and me? Tonight we celebrate and have a party!"

• • •

I saw them go as they walked away, again busily talking and laughing, some singing. I was excited; I'd been right. Everything had changed since yesterday. It seemed that the people themselves had changed now that they'd come in contact with the Creator. I couldn't wait to see how things would go from here.

10

REBELLION

IT ISN'T ALWAYS CONVENIENT to be part of a carbon atom. You're never alone and always have someone to talk to—that's a nice extra—and you can be fairly certain that you'll have an extremely long life. But you don't have a say about where you are taken. You can't decide to go somewhere; you're totally dependent on the mineral or organism that you're part of. You can sit for millions of years hidden deep within a rock, be buried for millennia under polar ice, or wander through the seas. You can be trapped in a tree trunk and then end up in the ground for years, where there's little outside interaction.

Of course, I wanted to know how things would continue to go for Womuntu and Maisha and the people in their tribe. I wanted to go with them to see the forbidden well. I wanted to see how they'd gather the fish from their nets in the river, how they would, together, prepare for the party, build a fire, and sing and dance for the Creator. How the Creator would perhaps appear again from out of the darkness, sit down between them around the fire, and talk to them as a father to his children. I wanted to know how they would shape their society and look after the plants and animals as the Creator intended them to. I wanted to see what would happen when they went to the holy well in a grand procession and drank its water.

Unfortunately, it's difficult to move around when you've become a part of a fence intended to keep deer and pigs out of a vegetable garden. In the evening we heard the sound of drums far away, but we didn't see anything. The next morning we saw men working in the vegetable garden. They laughed, joked with each other, and helped each other. Women and children came along to gather the harvest; they sang and thanked the Creator.

• • •

Some time passed. At dusk we saw deer come close and disappear again, and small groups of swine came passing by. Rain came and went. The gardeners sowed and planted; people weeded, chatted, and laughed. One morning—it had been rather noisy in the village the night before—part of the fence was overturned by a sow. Pigs pushed their way into the vegetable garden to look for insects and bulbs.

Soon people came along. They weren't happy, and they didn't laugh about what had happened. They were angry.

"Didn't I tell you to check the fence?" an older man said. "Why didn't you do it? Just look, they've dug up everything."

"Yes, and whose fault is that?" replied his friend, who seemed a lot younger.

"What do you think? If you'd repaired that fence, nothing would've happened."

"If you'd taken the trouble to make a solid fence, the hogs wouldn't have been able to walk right over it."

"If the Creator hadn't made those rotten beasts, we wouldn't have needed any fence at all."

The voices disappeared.

"This is strange," said Ensis.

"Very strange," I responded. "Are they blaming the Creator for something that they themselves did wrong?"

"I find that peculiar," said Aris. "We probably misunderstood them."

A couple of days later we saw two men fight each other. From the insults they snarled at each other, I understood that the disagreement had something to do with the harvest.

Of course, we knew that people, just as other animals, could be harsh to each other. We could see how they could pick a fight, how they could wound and kill each other. But all of that was before they'd met the Creator, before they'd chosen the new way—the way of love, peace, and harmony. Was this a kind of game that I didn't understand? I'd seen young predatory cats do something like it: the small ones would bite each other, not to hurt each other but to learn how to hunt. But the men became rather rough, and they were hurting each other. Eventually one of the men remained motionless on the ground, and the other man ran away.

"Is he dead?" Aris wondered aloud.

He was dead, we concluded, and I was totally confused.

"Didn't the Creator have a completely different society in mind?" I asked.

"It almost seems as if they've turned back time," said Aris. "This is how it was before."

"No," said Solon, "this is worse, much worse. Because they now know how things can be done differently, don't they? I don't understand what's happening, but this can't be the intent."

"Well, haven't you heard?" asked Airon, a proton in a molecule that'd landed near us during the fight. "You didn't think that things could remain without consequences?"

"What are you talking about?" I asked.

"What? You've heard about the rebellion, haven't you?"

I was shocked. "Rebellion?"

"Against the Creator."

"What?"

Airon told me what had happened.

• • •

During a meeting of the villagers, an older man named Mdala stood up to speak. "Dear leaders," he said, "honored Womuntu and Maisha, I want to ask you something."

"Speak, Father Mdala," said Maisha.

"Is it true that the Creator has said that we're not allowed to drink well water?"

Maisha smiled, "No, no, don't worry about that. We may drink from all the wells, only not from the well at the foot of the hill, east of the village, as you know. We may not drink the water from that well, we may not even use it to water the garden. The Creator has forbidden us to do that."

"There are rumors," Mdala said, "that those words are not correct. They say that you have forbidden us to use the water from that well because you know the secret power that it has."

Maisha looked at him with surprise. "That's not true at all," she replied. "Who said something like that? It has to be someone who does not know the truth yet. Bring him to us, then we can tell him everything."

"Wait just a moment, dear," Womuntu interrupted her. "Tell us more, father Mdala. I first want to know more about this rumor."

The old man bowed his head for a moment and then continued. "The Creator has given this well a strong healing power, it has been said. But the two of you don't want to share this power with us."

Womuntu was quiet for a moment, and then shook his head. "Thank you, Father Mdala. Thank you that you have shared this with us. What you have heard is not correct. We have shared with you what we have heard ourselves from the Creator. That we may not drink the water from that well, because that would break the bond that we have with him."

A woman raised her hand. “It is possible that the Creator has said this,” she said. “But that may be because he knows the secret power of the well and doesn’t want to share it with us.”

• • •

I so wanted to tell Airon to stop talking at that moment. I felt a great fear rising within me that I’d experienced only once before in my life—just before our star was about to collapse. I didn’t want to hear this story. I didn’t want to know how it would end, but Airon didn’t notice my anxiety and continued.

• • •

A wrinkle appeared in Womuntu’s brow. “The Creator knows all the secrets,” he said. “He tells us all we need to know, and that is enough.”

“We can ask him about it tonight,” said Maisha.

Mdala shook his head. “If it’s a secret, he won’t tell us,” he said. “He has an interest in concealing it.”

The meeting became tense. “How do we know we can trust you?” another woman challenged.

“How do we know whether we can trust the Creator?” Mdala asked. “Look at this boy here.” He pointed to Ardhi. “He drank from the well; he has admitted that himself. And . . . did he die? No. In fact, do you remember that he cut himself in the arm last week when his knife slipped? Do you remember how the wound bled? Ardhi, come here and let’s see your arm.”

The boy who’d earlier confessed that he’d drunk from the well came slowly to the front. With obvious reluctance he showed his upper arm.

“See!” Mdala said triumphantly. “The wound is closed and virtually healed. There’s just a small scar left.”

"Yes, and do you know why that is?" said a woman with a toddler at her breast. "We closed the wound and put medicinal leaves on it."

"The Creator has blessed our acts," the elder sitting next to her said.

Mdala ignored them. "This boy is healed," he shouted with a loud voice, "because he drank from the well the Creator has forbidden. That has saved him. It has caused his wound to heal so quickly that there's almost nothing left to see."

He turned again to Womuntu and Maisha. "I can't understand why we wouldn't be allowed to drink from a well with healing powers. That means that the Creator would withhold something from us that'd be good for us. He isn't like that, is he?"

Womuntu and Maisha looked at each other. "No," said Maisha, "he's not like that. He always wants the best for us."

"Forgive me if I didn't choose my words well," continued Mdala. "I don't want to accuse you of deception. Perhaps you've misunderstood his command. Perhaps the forbidden well is just the well from which we *should* drink. Hasn't this well been made by the Creator too? How can something that's been made by the Creator not be good for us?"

"That's true," admitted Womuntu. "Everything he's made is very good, including the water at the foot of the hill. Nevertheless, he does not want us to drink from it."

Mdala put the tips of his fingers together. "Could it be," he began, "that the Creator is putting us to the test in this? That he wants to see whether we are infant children or beings on his level who can decide for themselves whether something is good or not?"

"Let's discuss this with him ourselves tonight," Maisha proposed.

"But then we merely continue as we are," Mdala countered. "Then we continue to act like small children, like toddlers clasped to the breast of their mothers, even though they can easily feed themselves. Like children who cannot come to adulthood and look after their own responsibilities."

Womuntu frowned. "I do think Mdala may have a point. The Creator wants us to grow in faith, in love. How can we show that we are doing this? By making choices ourselves. Imagine it, Maisha, he comes to visit us tonight, and we can tell him we've taken this step. That we've come to adulthood. How happy it would make him. Maybe this was his intention all along."

Maisha was quiet for a moment. "I must say that it's a beautiful well," she admitted. A smile lit up her face and she sat up a little straighter. "With cool, clear water from the hills . . ."

"You say it yourself," Mdala interrupted her, "a beautiful well. Delicious water. Why would the Creator want to keep that for himself?"

"Exactly. Why?"

"That would be unfair," someone from the crowd called out.

"Because the water has secret powers?" cried another.

The noise in the meeting was rising. People began to discuss things among themselves and to call out to each other.

"The well lies in our territory, so it's our well."

"Yes, but the Creator said . . ."

"Now what! 'The Creator said?' Doesn't he want our sick to be healed?"

"I don't know about that, but I do know we can trust him."

"How do you know? Maybe the water will make us as powerful as he is, and he just doesn't want that."

"I agree. He doesn't want us to discover his secret!"

• • •

"I don't know how it is with you," said Womuntu, turning to Maisha, "but the more they talk about the well, the thirstier I become."

"Me too," said Maisha. "That's the strange thing. I want to drink that water more and more. Why don't we just go there, you and me?"

"Just imagine that the water does in fact indeed possess a secret power . . ."

"I'm getting that feeling more and more, that there's a superpower in that well that'd give us access to knowledge that only the Creator has, when we drink the water."

"Yes! And that power and wisdom . . . don't we just need it to be able to do our work?"

• • •

"Why's it taking so long?" someone yelled.

"If you don't give it to us, we'll go and get it ourselves."

"Water, water," a few men began to yell. They jumped up and pumped their fists in the air. "Water, water!" The children jumped along, and then, the women. Everyone jumped, danced, and chanted. "Water, water!"

Womuntu climbed onto a boulder. He raised his hand.

"Fathers and mothers, brothers and sisters," he shouted, and waited until it finally became quiet. "Maisha and I have talked about it. We're going to the well together, and we'll drink from it. And when we've had our drink, you may drink too. We won't keep our divine wisdom and power for ourselves. We'll share it with you."

"Thank you Womuntu and Maisha," Mdala exclaimed above the cheers. "Thank you on behalf of everyone. You've made a wise choice. The Creator will be proud of you."

11

DEVASTATION

I WAS IN SHOCK. This was a story I didn't want to hear. Still, I knew that I couldn't just close my ears. I had to know what'd occurred.

"And then?" I asked anxiously.

Airon told me what'd happened afterward. How the people had run to the well with Womuntu and Maisha, how the strongest men had taken the stones away, how people had shoved each other to be able to drink from the well, and how they'd pushed and pulled each other. How people had been in a festive mood as they returned and partied late into the night. And how, on the next day, the first cracks seemed to appear in their happiness. The euphoria of the previous evening had disappeared, and the atmosphere was totally changed. People were short with each other, and there was little conversation, certainly not about the events of the evening before. The mood also affected the children. They whined, slept badly, and bickered with each other during the day.

Maisha and Womuntu had a meeting with the Creator, from which they returned dejected. Afterward they got into a terrible quarrel that resulted in Womuntu taking his dugout and sailing off alone. When he returned the next day, he discovered that Maisha had put herself forward as the new tribal chief and had gotten the support of more than half the villagers.

When an older woman died a few days later, Womuntu was summoned from his hut and called to account by the angry family.

"What did you do with the water?" cried one of daughters of the woman who had died. She, like rest of the family members, had smeared her face with white clay as a sign of mourning, and she grieved loudly.

"You said it was good," accused another. "You said this water would heal our sick."

"He pronounced a curse over it," yelled a man who wasn't related to the family. "My wife saw him when he, with that wife of his, went to the well by the hill early in the morning. They've cast a spell over it."

Womuntu tried to say something, but he didn't get a chance. "Or it's his God. His God has spoken a curse over it."

"We trusted you, Womuntu, we chose to be on your side, but it only brought us bad luck."

"Your God brought us misfortune."

"Justice has to be done. A life for a life. We demand sixty goats in return for the life of our mother."

• • •

We were upset by Airon's story.

"Why?" Ensis cried. "Why have they done this? Why did they violate the commandment? How could they be so stupid!"

I wondered about this too. The people who had been so fortunate that they—the only creature that'd been addressed by the Creator—had been allowed to get to know him as their father and friend; who could talk to him; for whom he wanted to make the earth into a home where they could live in peace, on the way toward an existence without illness, without pain, without death. . . . Had they really knowingly broken that bond? How was it possible?

“It’s not just the human beings,” said Airon. “It seems there are other powers involved, invisible powers that draw the people.”

“There’s a war going on,” one of his friends added. “We don’t know the ins and outs of it, but the Creator seems to have a dark adversary who is out to take control into his own hands. He’s mobilized a large army, and he knew exactly how he could strike a blow at the heart of the Creator: by obstructing his favorite project, by sowing discord and getting the people to support him.”

“And that without the people realizing it,” said Airon. “The people haven’t noticed a thing. Without realizing it, they have gone over to the side of the opponent.”

“Everyone?” I asked.

“Everyone. As far as we can understand it, evil has spread very rapidly. It was like an infectious virus. There was no human left who wasn’t affected by it.”

“Has the Creator’s opponent poisoned the well?” I asked.

“No, as far as we know, there was nothing wrong with the water. It was that the people chose to drink from the well, that they broke the only command the Creator had given them. With that they pushed him out of their lives. There is no return from that,” Airon said somberly.

“No, really?” I asked.

“Billions of years . . .” Ensis sighed.

She didn’t finish her sentence, but we all knew what she wanted to say.

Billions of years the Creator had worked to prepare for this project, with so much love, so much creativity, so much patience, so much attention. He’d been so happy to share his love with this creature. He’d spoken to them. He cared for them as a mother for her children. But they’d broken their tie with the Creator. They chose for themselves.

Whoever the unknown opponent was, whatever his role had been, one thing was clear to me: he had torpedoed the project of the Creator and had turned the people to his side.

"And now?" I asked. "What's going to happen now?"

"A meteorite," predicted Aris. "Maybe even a bit bigger. All life will be gone."

"A new start, somewhere else in the universe," said Solon. "That's a possibility."

"A completely new universe," said Axion.

"Or nothing anymore," I said. "The Creator doesn't need anything to make himself complete. Or to be happy."

"Well," Airon resumed. "That's the strange thing. It seems the Creator has decided to continue this project nevertheless. He's spoken with Womuntu and Maisha, I've heard. He's deeply disappointed and devastated about what they've done, but he's become very attached to the human beings, it seems. He doesn't want them to lose themselves to tragedy. He's looking for a way to win them back to himself."

12

DOUBT

I KEPT A CLOSE EYE on my surroundings to see what would happen from the fence that we were part of. I wondered what the Creator was going to do now. I couldn't imagine that he was going to leave it at that. That day no meteorite came falling from space to put an end to life on Earth, our star didn't explode, the laws of nature simply remained in force. The Earth continued to orbit around the sun as if nothing had happened. People continued to come to the vegetable garden; the fence was fixed; and gardeners plowed, planted, weeded, and harvested. Still, everything was different; at least, it seemed that way to me.

The Creator no longer stopped by. That was the first thing I noticed. I wondered if he'd turned his back on people. Perhaps he'd forgotten them.

I also noticed—Airon had already talked about it—that the mood among the villagers had changed entirely. People that I'd seen working together in as friends now seemed to be annoyed with each other. They gossiped about one another and argued. They complained about the level of the water in the river being so low, about the hard ground, the nettles and thistles, and the insects. They grumbled over the unfair distribution of the work (everyone thought they had too much), about what a shame it was that Womuntu and Maisha had taken the largest part of the harvest (while it was already so disappointing), and that they had also demanded the best part of the deer, even though Womuntu hadn't been present when it'd been brought down.

"I just don't understand what makes the atmosphere so different," Aris sighed. "I mean, before the Creator came to earth, they also had to work; they argued, they fought, and sometimes people were even killed. Really, it's just as it was before."

"No, it's different," countered Solon. "When you don't know something, you don't miss it. Before the Creator came, everyone lived their life as it came. They couldn't do it any differently. But now they know how it could have been. Do you still remember what it was like to live in the dark, right after the very first beginning, when time was just born?"

I thought about it. "I didn't find it a problem," I said. "I just existed. I had no idea that there was such a thing as light."

"And then came the light," Aris replied. "Do you still remember that, Pro?"

Of course, I did. I remembered as if it had happened yesterday. The surprise. The delight. It began with short, weak flashes of light that penetrated the fog. It was incomparable with how we would perceive light later. It was a totally new dimension. The creation appeared to be so much richer, so much more beautiful than I could ever have imagined.

"After that, it disappeared again," said Ensis.

"Oh, yes," remembered Aris. "That was horrible."

She was right. I hadn't experienced the first darkness as negative. However, during the second darkness I felt a constant emptiness. I thought we were finished, that the project of the Creator had ended in failure. In the long time that followed, I thought about the light and longed to be able to see it again. Once you've experienced the light, you can't be happy with darkness anymore.

"And the colors," said Solon. "Do you still remember when we first discovered that the Creator had enclosed colors within the light? What a surprise that was? And now we can't imagine missing all that variety. I think it is a bit like that for people too. They've experienced the closeness of the Creator. It was as if the light turned on for them. They tasted it. They saw how beautiful it was. They felt how happy it made them. And then they extinguished the light, themselves. . . . Of course, things are no longer as they were before. It's not possible. They will always long for the way it was."

"Is there still hope?" I asked.

"According to what I've heard, there is still hope," said Solon. "But we have no clue how. Or what. Or when."

"How long was it before the light reappeared?" asked Aris.

"It is difficult to say," replied Solon. "But if you were to figure in sun time, it would be about four hundred million years."

• • •

I was finding the daytime temperature to be too warm. Not as it was long ago—during the time when the star was our home, for compared to that it was rather cool—but according to earthly concepts it was hot. It hadn't rained for months. The soil in the garden dried out, the clay had deep cracks, and the crops hardly grew any longer. We heard that this was so because the holy well was drying up. Most of the people left the village and moved away. The fence again collapsed. A small boy took the bamboo stick to which we belonged to his home, and that evening we were thrown on the fire, apparently because a rat was going to be roasted.

"Here we go again," murmured Aris when the fire was lit.

It became warmer and warmer. The wood in the fire fell apart, and in a swirl of heat and smoke in the air we ended up in a CO_2 molecule, were taken along in the evening breeze, floated higher and higher, and landed in the desert some days later.

13

RESTART

THOUSANDS OF YEARS LATER we were part of a sheep, a complete world in itself, with billions of cells, each cell a crowded city. We lived in a cell in the outermost part of the skin. One day the sheep was slaughtered, the intestines were removed, and the hide was pulled off and scraped clean. The hide was tanned in a mixture of water and oak bark and hung out to dry in the warm breeze. Then we were taken, along with other hides, to the walled-in courtyard of a large stone house and thrown onto a pile.

"Where are we?" I asked.

"They call this place Ur," said a proton from a dust particle that came floating by.

"That doesn't tell me anything," I said. "Where's that?"

I got no answer. My attention was drawn to a man and a boy who walked together into the courtyard.

"So many sheepskins!" cried the boy. "What are we going to do with them, Uncle Abram? Is this going to be a new carpet?"

"No," said Abram. "We're going to have them made into tents. Big tents."

"What do we need tents for? Don't we have a house?"

"We're going to move. I'm organizing everything. I don't know exactly where we are going yet, but I have become convinced that we will be shown the way to a better place. Maybe we'll have a house there again, but while we're on our way, we'll sleep in tents."

The boy frowned. "Are we going to leave Ur? That's dangerous, isn't it? Can Sin still help us when we're in that other land? And the other gods?"

"We'll take them with us, Lot. Don't worry. Their images are powerful, and they'll protect us on the way."

"Are all of us going?" the boy asked. "With the whole family? Is grandfather Arpachshad coming too? I'm going to ask him." He walked inside. "Grandpa! Are you coming with us to Canaan?"

"He's sleeping, Lot," said a woman's voice. "Let him be. Oh no, that's so annoying. Now you've done it. He's awake."

"Ah, the days of yesteryear," said a trembling elderly voice. "The days my father told us about. The days of the big flood. Water, as far as the eye could see. It came up out of the earth. It came down from heaven. Forty days, forty nights. Sea everywhere. Not one human being that survived. Not one. Only my parents and the animals they had with them on their boat."

"Are you telling the stories of long ago again, Grandfather?" asked the woman. Her voice was gentle. "Here, drink some wine. Lot, please go outside."

The voice of the old man rose. "'And when you see the bow in the clouds,' he said, 'then think of what I have promised today. Never again shall . . .'"

"Shh, don't worry, Grandfather. That's not good for you. You're safe. Do you want something to eat, perhaps? I have bread, soaked in goat's milk. I'll help you. Lot, away with you, and close the door behind you."

The boy stomped into the courtyard looking annoyed. "My mom doesn't even allow me to ask him a question!"

"I think she's right, Lot," said Abram. "My grandfather Arpachshad lives in the past. He's not going to come with us. He's going to stay here with some servants who'll take care of him."

"But . . . why?"

"He's too weak, Lot, and much too old. He's ready for his last journey. Not long from now he will go on his way to the country from which no one returns—the underworld, where every person, rich or poor, good or evil, awaits the same dark destiny."

• • •

"Wait a minute," I said, when Abram and his nephew had left the courtyard. We were still lying in the sun against the little wall. "Is the man saying that there's a god with the name Sin, who can protect him?"

"That's what he was talking about," said Aris, "and about the images that he's going to take along on his journey."

"Remarkable," I said.

"I didn't know that they call the Creator Sin nowadays."

"If you ask me," said Solon, "they have no idea who the Creator is. They've invented their own gods."

• • •

A couple of months later, now part of a tent, we were loaded onto the back of a camel and taken on a months-long journey westward. During the days we saw little because we were bundled up and packed. We felt only the moving and shaking of the camel transporting us. But when we were unpacked and set up, we saw endless sandy plains, rocks, hills, and once in a while a green oasis where we would stay for a couple of days. We saw the sun set behind the hills, and at night we saw thousands of stars move on their paths through the heavens. At dawn we saw the moon and stars fade, and we marveled at how the Creator brought color back into the world every morning.

Finally we arrived in a city called Haran. Abram's father decided that a short rest would do him good. Then he found the place so pleasant—even better than Ur, he said to himself—that he wanted to stay there to live. My friends and I ended up in a storeroom, and we must have been there for years.

We saw daylight again when Terah had died. Abram's hair had turned gray, his wife Sarai's face had become wrinkled, and his cousin Lot had become a grown man.

"My husband has pulled out the old tent?" we heard Sarai remark. "May I ask you why, my lord?"

"We have to go again," said Abram. "We're going to travel again. Sarai, Lot, I have to tell you something. Last night the Lord of heaven and earth spoke with me."

"Sin?" said Sarai. She sounded shocked.

"No, not Sin. We've been misled, Sarai. Sin is not a god. Shamash and Ishtar, they aren't gods. We and our parents put our trust in stone and wood, instead of in a living God." And with these words, Abram grabbed the stones that he'd mistakenly seen as gods his whole life and threw them down so hard that they shattered into pieces on the ground.

"No, Uncle Abram!" Lot shouted. "No! What's gotten into you? The punishment of Sin is going to be terrible."

"Sin can neither hear nor see, Lot. He can't reward or punish; he can't do anything. He's just a dumb piece of stone, just dust as we are. Listen. The Almighty, the God who made heaven and earth, spoke with me last night. We've turned our backs on him, just as our ancestors did, but he sought us out. He said that he loves us and wants to make a new start. With us. With you and me, Sarai. And you can be part of this too, Lot. The Lord wants us to leave Haran and continue our journey again, together with him. He wants to give us a country, and not only a country. He wants to bless us and fill that land with our descendants."

"But Abram," Sarai said softly, "you know that Sin has never given us a child. I am too old, you know that too."

"Sin has no power, Sarai," Abram answered forcefully. "The God of heaven and earth has all the power, even to give us old people a son. He has other plans besides that. He's not only concerned about us but about the whole world and all people. He hasn't explained his plans exactly to me, but I understand that through our family he ultimately wants to reach out to all people with his love. Through us he wants to reconnect with human beings. It's unbelievable!"

"But uncle," Lot said. "Didn't Grandfather have good reasons for discontinuing our journey? It is too dangerous! Didn't you, yourself, tell us the stories about caravans being robbed? About cattle being stolen, women being taken, and men being murdered? And we're doing so well here!"

"The Lord will protect us," said Abram. "He has assured me of that. Whoever treats us well will be rewarded. Whoever wants to harm us will be stopped. He shall not escape the Lord's punishment. I'm going to talk to the servants and tell them to get everything ready for our departure."

● ● ●

"Didn't I say it!" Aris shouted triumphantly. "Didn't I say that the Creator wouldn't abandon the people!"

"I can't remember that," I said. "I think you said he was going to wipe them out with a meteorite."

"Nonsense," said Aris. "Where would you get an idea like that?"

"This is good news in any case," said Solon. "It appears that the Creator has made a restart with his project."

14

LIBERATION

Soon we became accustomed to the rhythm of traveling, being set up, and serving as the place for the people to sleep.

"What's going to happen to us when we're in Canaan?" I wondered aloud.

"What do you think?" said Ensis. "We're going to be discarded, of course. If they can choose, they're going to choose a house, not a tent."

But there was no choice. Canaan wasn't empty; it was a land occupied by people, subdivided among the inhabitants. The people living there were not inclined to give up their land and houses and move elsewhere. Abraham and Sarah, as they were now called for some reason, were forced to continue to use us and the other tents, and to camp their entire lives, here a while and there a while.

The son that had been promised didn't come either. I saw with some anxiety how Abraham and Sarah were becoming older and older, and I noticed that Sarah had given up all hope. "I'm no longer needed and totally worn out," I heard her say one morning. "Totally worn out. That husband of mine used all kinds of sweet talk to get me to come to this country, but I shouldn't have believed him. It all turned out to be nothing. I no longer have a house or land, and I have no child."

• • •

“What a faith,” Ensis said scornfully.

“She’s right, it seems to me,” I said. “I do wonder, by the way, if her husband still believes the words of the Creator. It seems as if he’s put all his hopes on the son he fathered with the slave woman. I get that too.”

My friends and I had been part of all sorts of cells, including egg cells that were necessary to begin a new life. We knew that after reaching a certain age a woman no longer produced this kind of cell and her body could no longer sustain them. It was impossible for Sarah to become pregnant.

But one morning I heard a strange sound coming from the women’s tent, a baby crying. Abraham’s wife had given birth to a son. I couldn’t understand it at all.

“How can this be?’ I asked.

“The Creator promised them a child,” Solon said. “He made the laws. It wouldn’t be a problem for him to do things differently one time, to see to it that an infertile woman would become fertile again.”

“It seems to be of great importance that this child be born,” Aris observed.

“That plan,” said Solon. “That plan the Creator still had to save the people . . . Would it have to do with this?”

A feverish excitement took hold of us. We kept a close watch on the child, as much as we could. We saw him as a child in his old mother’s arms. We saw him crawl around outside the tent. We saw his first feeble steps. We saw how he was pushed around by his half brother. We heard him babble and later talk. We heard him laugh and sometimes cry. Was the Creator going to make a new start with this child? Would he visit this boy to talk with him as he used to do with Womuntu?

• • •

I couldn't keep up with the time; the years flew by so quickly. The small boy, who had been given the name Isaac, grew up and became an adult. He married a woman who also couldn't have children. The family line threatened to end there. Isaac begged the Creator for help. A year later his wife, Rebecca, gave birth to twins, Esau and Jacob.

None of this was world news. No one suspected that something special was happening in this small nomad family. But for the first time in a long time, I heard people tell their children about the Creator again, heard them sing for him and speak to him. Apparently they understood that he was present and involved in their lives, that he could hear them, even if they couldn't hear or see him.

I saw how they stacked stones on top of each other on certain occasions, to make a kind of elevation on which they burned grain or meat. It was their way of giving an offering to the Creator to thank him for something.

Unfortunately, all kinds of things also went wrong in this family, especially when the twins became older. I saw Jacob and Esau argue and fight. I saw how Isaac and his wife turned against each other, deceived each other, and ignored the Creator's directions. In the end, Isaac died, old and blind, after a life that ultimately differed little from that of his ancestors. Just like the people before him, he hadn't always been able to choose the right way.

For some years it seemed that the whole project of the Creator would come to an end—a long period of drought in Canaan caused the crops to fail. There was less and less to eat. Children and pregnant women were at risk of dying from malnutrition. Jacob and his family were also affected by the famine. I was worried. This could turn out to be the end of the family.

Then they were offered refuge in Egypt. Traveling again! A new adventure! We were taken apart, packed, and loaded on donkeys. Together with old Jacob and his extensive family, we traveled to the eastern part of the Nile delta in Egypt.

Once we were there, we and our companion molecules were dismissed, dusted off, folded up, and put in a shed. Finally the descendants of Abraham could live in houses. For me and my friends, however, it meant that we were largely deprived of news. Our shed was dark and musty. There was nothing to be seen, heard, or experienced. We had to amuse ourselves as best we could.

• • •

It must have been hundreds of years later when, one evening, we were pulled out again. It was dark, and we were thrown onto the dirt floor. We found ourselves in a simple home with walls of clay mixed with straw. There was a nervous excitement in the room. I saw a mother rolling up a sleeping mat and a six-year-old boy with a piece of bread in his hand.

"Why is the bread so flat, Mom?" he grumbled.

"Because there's no time to let it rise, Simi. I explained that to you. Keep eating. Haven't you put your shoes on yet?"

"I don't like flat bread."

"You'd better eat it anyway, because we're going to leave right away, to go back to Canaan—the land your grandma told you about so many times. The land where everyone drinks milk every day, where you can dip your bread into honey and drink wine with every meal. The land that the Lord, may his name be praised, promised to our ancestors."

"Will Pharaoh allow that? Because he is very cruel, right? You yourselves told me so."

"The Lord is going to ensure that we're allowed to leave. He's going to set us free. Keep eating."

I watched as a man came into the house and lifted a package from the floor.

"Hosea," said the woman. "How are we going to manage with that tent? It's way too heavy for us, and we have to take the rest of our belongings and food for on the way."

"No problem, we have an ox."

"What're you saying? Since when do we have an ox?"

The man laughed. "Since about a half an hour ago. We got it from the people across the way."

"What?"

"Look, they've given us all their necklaces and rings to take along, with the request that we would leave, please. They're scared to death of us after what happened last night. Pharaoh himself begged us to leave."

We left in the middle of the night strapped to the back of an ox, together with a large group of people and cattle. Gradually we found out what had happened. Egypt, the land that had been a place of safety for us and where the family of Jacob was able to grow into a people, had become a prison. Pharaoh had come to see the refugees as a threat, and he'd taken harsh measures. He'd forced the men to serve him as slaves and had ordered all newborn boys to be killed.

Without us realizing it, the family of Abraham had once again been threatened with extinction. They narrowly escaped, as we heard, by the intervention of the Creator himself.

• • •

During the journey the Creator protected his people against enemy attacks, made sure that they had food and drink, and gave them a basic sort of law. They were rules that'd been so obvious to Womuntu and Maisha that they hadn't needed to be recorded: to love the Creator and fellow human beings, and to care for their environment.

"Strange, really," mused Ensis. "The rules to take care of the weak . . . that doesn't seem to me to be very efficient. Why not just prefer the right of the strongest? Why doesn't the Creator say that whoever can adapt themselves the best, survives? That is the tried-and-tested method. We've known for millions of years that it works."

"Perhaps because the Creator has another sort of society in mind?" wondered Solon.

"Yes, Ensis," I replied. "This is all about love. Of course," I razzed my longtime friend, "as a neutron, that's not something you would easily understand."

15

IMPASSE

IT TOOK FORTY YEARS before we were back in Canaan—much longer than it needed to be, if you ask me, because it was not that far away, really. Sometimes I worried that these people would just start to live in the desert permanently. But finally we crossed the river at the boundary to Canaan under the leadership of Hosea, with whom we had once left Egypt, and who, for reasons unknown to me, was now called Joshua by everyone. In the centuries that followed, the family came to possess an ever-larger area in Canaan. That didn't always happen in the gentlest way.

Now that we were back in Canaan, would someone arise who could see to it that life would again become as the Creator had originally intended it to be? To be honest, I didn't have much hope for that, when I heard how people got along with each other. How quickly they forgot the Creator and were persuaded to pay homage to a stone or piece of wood. How a poor woman with children had to sleep outside without a blanket because she couldn't pay her debts, and had to give her only possession—a blanket—as security. How judges were bribed.

My friends thought I shouldn't be so pessimistic. They told me stories about people who honored the Creator, who tried to live according to the basic law that he had given, who thanked the Creator for the good in their lives, and who sang for him.

Of course, it may well be that I missed something. When you're in the ground, you don't see much of what's happening. And when you're a component of a sheep intestine—as we were some five hundred years after arriving in Canaan—you're not always fully informed. But when the sheep in which we found ourselves was slaughtered, we were—with intestines and the rest rinsed out—scraped clean and promoted to being a musical string. We were strung on a harp and played by a man called David, who turned out to be the king of Israel.

One evening he took a us along to the roof terrace of his home, an enormous, luxurious building with thick outer walls that provided protection, and with interior walls and floors made of cedar wood. It was a warm, cloudless night with a half moon. Far above us were the stars, unreachably far away, the dreams of a time long ago. The king strummed the strings and sang a song.

Lord, our God,
How great is your power.
I look at the heavens
That you have made,
the moon and the stars
fashioned by you.
How great you are, Lord,
so far and yet so near.
I cannot fathom
that you look after us
and even love us,
that you give us a task
and make us kings.
The care for your earth,
the animals and forests,
the water, the air,
you entrust to us.

Evil will be overcome.
How great is your power!

• • •

"Do you hear what he's saying?" Ensis cried jubilantly. "This man. He almost sounds like a new Womuntu!"

"This is indeed special," said Solon. "This is someone who sees the Creator and honors him."

"Very beautiful," I said, "but I still have to see how this turns out. I wonder if he can get the people to follow him."

"Not so negative, Pro," said Ensis.

"No, Pro," said Aris. "Not so pessimistic. If anyone is going to succeed, it's this man. He's king, after all!"

"Mind you," said another proton, "the Creator has told him that he has chosen him and that someone from his family will always sit on the throne."

"You should've heard all the things he does," said someone else. "Of course, you wouldn't be aware of that because you were a sheep intestine."

"Which was an honorable function," Solon interrupted. "A task that was entrusted to us."

"Of course, I don't want to take anything away from that. But we've seen how this man has made a new dwelling for the Creator. He's reinstituted thank offerings being offered and that the people could learn to know the Creator once again. 'The man after God's heart,' he's being called."

"See," Aris said triumphantly. "I knew it. This is him. I have good intuition for this kind of thing."

• • •

But a couple of years later, the man after God's heart seduced a married woman, made her pregnant, and had her husband killed to prevent his misstep from becoming known.

"See," I said. "I just knew it."

"Okay, okay," said Ensis. "This went against all the intentions of the Creator. But David regrets it."

"Womuntu and Maisha also had their regrets."

"Nevertheless, I think that Pro is right," said Solon. "It's a pity, but this may not be the man that brings the people back to the Creator. Who would it be then?"

That's what I kept wondering during the years that followed, when the son of David began to worship other gods. When the country split into two parts. When I saw rich people exploiting the poor. When I observed that the basic laws of the Creator were being forgotten time after time. When war broke out and thousands were taken away as exiles to Mesopotamia. Fewer and fewer people remained behind from the people with whom the Creator had wanted to continue his special relationship.

It reminded me of the time, more than thirteen billion years ago, when it had become dark, cold, and lonely in the universe—a dire situation with hardly any prospects. Very infrequently, there was a flash of hope. Very rarely there was a voice of a man or a woman that passed on the words of the Creator, words of hope, salvation, and peace. There was news about a king from David's family who was going to make sure that all the people in the world would live in peace. There was an announcement of a star that was going to shine.

When was this going to happen? No one told us. One thing was clear to me by now—the Creator himself would have to intervene. The people he had made, even the best among them, weren't capable of bringing this about.

16

AVATAR

"I'M AFRAID WE'RE ON A DEAD-END ROAD," I said, about a thousand years later. We found ourselves in a spiderweb in the corner of an animal stall. The light was dim, and the stall smelled distinctly of the farm animals that were kept there.

"What do you mean?" Solon asked.

"I don't think the Creator has assessed the situation well," I said. "Or maybe those messengers of his have misunderstood his words—those preachers who were critical but always predicted a happy end. This, however, isn't going to turn out well. It simply can't. After a hundred tries, we can draw that conclusion, I'd say. What is left of the people he was going to make a new start with? Not much after all those deportations."

"Some people have come back," Ensis said.

"Just a few," I said. "A small portion of the people who were deported. The temple they built for the Creator is destroyed. The one built to take its place may recently have been refurbished, but it can't compare to what was there before. Besides, a descendant of David was always going to sit on his throne, right? That just hasn't happened. The Roman emperor is the ruler now."

"Not entirely though," said Aris. "There is a king in Jerusalem named Herod, if I'm not mistaken."

"But he's not from the family of David," said Solon. "He's from the family of Esau. You have a point, Pro. I can't see how that's in keeping with the promise that a descendant of David would always sit on the throne. The way I see it, it's not going to work out. Although we only hear snippets, so it's possible, of course, that we've misunderstood something."

• • •

"Yes, I'm sorry," a man's voice said apologetically, opening the door of the room. "We really haven't got anything else at the moment, and the guest room is occupied. We're not so crowded here, normally, but with this census of the people happening . . . I don't think your wife will want to give birth among the other guests anyway. Feel free to push some of the stuff aside."

"This is fine," said the young man who followed him. "At least it's warm and dry, right, Mary? We'll make it work."

The woman—a young girl, really—nodded. Her face was contorted with pain. Slowly she let herself down onto a bale of straw. Dust swirled in the air.

"The well is a long way out of town, but I'll have my wife put a bucket of water just outside here. If you need anything else, you can knock on our door upstairs." He closed the door.

I saw how the man looked around the place and pulled out an old feed trough from behind some farming supplies. "What do you think of this?" he asked as he put it in the middle of the stall. "I'll clean it out, put some straw in it. It'll be a good place for the baby, I think. Something is better than nothing."

The woman smiled. "Who would have thought it, Joseph," she said, "that the Son of the almighty God would be born in an animal stall like this. That a feed trough for sheep would be his place to sleep. I wonder if anyone is ever going to believe this."

"I wouldn't have believed it myself if the angel hadn't appeared to me," said the man. He sat down beside his wife and held her. "You, the mother of the Savior of Israel, the Son of

David, for whom we've waited so long. That we're allowed to experience this."

• • •

"What? What did they say?" It shocked me to realize what was happening. "Did you hear that too?" I asked. "Did that woman really say that she was going to become the mother of the almighty God?"

"She said something like that," agreed Ensis. "I heard it too."

"Well, people are so easy to deceive. This just cannot be," countered Aris. "The Creator doesn't have a son. She probably means that this baby is going to have a special position."

"But they were talking about an angel who told them about this!" I said. "The Creator could've sent one of his special messengers. If that's so, it must surely be true, right?"

"I don't know," said Solon. "I assume that this baby may be given a special position when he's grown, just like Womuntu. That he'll live as the Creator intended. That he'll be a new beginning. If so, this is a great day."

"Much greater than you think," called a neutron from somewhere else in the barn. She'd clearly traveled with the visitors, so she informed us, "These people really did have a visit from an angel, in fact from the highest representative of the Creator."

"Are you serious?" I responded skeptically.

"I mean it. They received a visit from Gabriel himself."

The neutron and her friends stumbled over their words in their hurry to tell the story.

"Seriously. From Gabriel himself."

"He just came to her house."

"And to his house too."

"He told them everything."

"Not everything . . . "

"No, of course not everything."

"Everything they had to know."

"But what did he tell them?" I interrupted them. "What did he say?"

"He said that the Creator himself would make sure that this woman was going to have a child."

"Oh, just a child," Aris said. "Together with her husband then."

"No! Without a man being involved."

"What?" I said bewildered. "How is that possible?"

"Yes, how is that possible!"

One of the neutrons urged her friends to be silent. "Can I squeeze in a word?" Then she started to speak, "We don't understand it. The girl didn't understand it either. We heard that the Creator used a clump of cells in this woman that grew in her to become a human being. That's all I know. In any case, they call that baby the Son of the Creator."

"His son?" I muttered to myself. The idea that the Creator might have a son would never have occurred to me. What surprised me most, however, was that the Creator would want to become a human being, to become part of his own creation. Wasn't that an irresponsible and big risk?

It was already deep into the night when the baby boy was born. To be honest, there was nothing unusual about him. This was a very ordinary human child who, after the birth howled lustily and only became silent when tightly wrapped in cloth and laid in the arms of his mother. This baby, this bundle of atoms, formed from stardust, could he really be the Son of the Creator? It was a bizarre idea, too fantastic for words. I knew the Creator felt very involved in his creation, particularly with people. I also knew that he'd long been busy with a plan to come to their rescue, but would he really go so far as to take on the rescue himself? Would he make himself one with his creation in order to rescue it? Was it possible that the Creator of the cosmos who carried this whole universe in his hand had stepped into his own creation in the person of his son? No, *stepped into* was not the right phrase; he was squeezed into it, screaming and struggling, unable to keep himself warm, talk, or look after himself. He was completely helpless; he wouldn't survive two days if there wasn't a woman to take care of him and feed him with milk from her own breasts.

If this was true, then this was the opposite of the Big Bang, as the Creator—who had brought forth the universe, space, and time from the smallest possible germ—laid aside his position of power to make himself incredibly small . . . to become a human baby. Really?

I couldn't comprehend it. We'd expected the rescue from the human side, an offspring of Abraham, of David, who would make a new start. A person who would receive a special assignment, who would do what Womuntu, Abraham, David, and all their descendants had not been able to do. Yet, this *was* a human. A tiny human being. A descendant of David even. But if what Joseph and Mary had said was true, the child wasn't only human but also the Son of the Creator himself.

• • •

After the initial euphoria I started to have some concerns. I didn't want to talk about them because it seemed inappropriate. But Aris articulated exactly what I was thinking.

"I don't know if I can comment on it," she said carefully. "But if all this is true . . . is this really such a good plan? Why would the Creator do such a thing? Put himself in a place where he is so dependent on people?"

"That's what I think too," said Ensis. "To be honest, I don't think this is wise. The Creator has seen what's become of his world, hasn't he? And what people have made of it? How few people there are that answer his love? Why didn't he wait until they went extinct, until a new species emerged with more intelligence? One that is not so foolish as to walk away from him?"

"I personally think he should've done away with the Milky Way and started somewhere else in another galaxy," continued Aris, who always looked for radical solutions. "Or he could start a new universe. That might be even better."

"It probably has to do with a dimension that we can't understand," said Solon, "with that love he's talked about so often. Of course he could've made a start somewhere else. But apparently he has given his heart to *Homo sapiens*. He loves them so much that he wants to share in their lives."

• • •

Down below us, we saw how the woman was busy with the baby as if she'd never done anything else. "Now, little boy," the woman said, and I saw how she wrapped an extra blanket around him. "There, nice and warm. This is better, isn't it?"

"Jesus," said her husband. "We're going to call you Jesus, little boy. Now look, Mary. He has a nose like yours. Shall I lay him in the feed trough? Then you can sleep for a little while."

"No," said his mother, "not yet. I want to keep him with me first. Do you want to drink, Jesus? Oh look, Joseph. He was hungry."

"Look at those little fingers."

"And those little nails."

"I'm going to make a carpenter out of you, Jesus. I am going to make you into a very good carpenter."

• • •

From our spiderweb we looked down on someone who (if it was true) was one with the Creator of heaven and earth. We saw a helpless baby who, right at this moment, was burping milk all over his mother. His greatness couldn't be seen.

"Okay," I said, "suppose he wants to live among people, and he wants to take over the task of Womuntu. Then why does he come in this form? Wouldn't it have been better for him to come as an adult, as a mighty emperor, or as a wise philosopher? No one will recognize him now, let alone be in awe of him."

• • •

Voices came from outside. A bang on the door. The little Jesus, who now lay in the feed trough, began to cry. He was so tightly wrapped that he could move neither his little arms nor legs—the normal reflex of a newborn monkey or baby when startled by an unexpected sound.

Joseph got up, walked to the door, and opened it a crack. "Sorry to disturb you," said a man's voice. "But . . . eh . . . there was an angel who said that this was where we had to be. Is that right?"

The door was opened further, and a group of rough-looking men came in.

"Hey, man, look. In the feed trough! He's lying there."

"This has to be him. The savior of the world that the angel spoke about."

The men fell on their knees in front of the feed trough and said a prayer of thanks. "Thank you, God of Israel. You have not forgotten us."

• • •

"And you thought he wouldn't be recognized, Proton?" chuckled Solon.

As soon as they were able, the family moved into the guest room of the farmhouse. I'd expected a stream of visitors to come to the baby, but that didn't happen. A single neighbor woman, a few small children who'd heard the crying and became curious. No one else.

But one evening, a while after the baby was born, the stall door opened again. Two men came in who, from the sound of their voices, were not from Israel. They talked to each other about where they could best let their camels rest.

"Not here," said the older of the two while he looked around. "This is way too small. We'll just leave them outside. Our masters won't be staying that long."

"Strange place, this," said the other, a man with a short beard. "I'd expected something grander for a king who was so important that his birth could be read in the heavens. I really wonder if we're in the right place. I mean, there aren't even any guards here. The masters were able to go straight through to upstairs. They didn't even have to request an audience."

The older man shrugged his shoulders. "They probably know what they're doing. They know more about the stars than we do."

There were footsteps that came down the stairs and enthusiastic voices. Then the sound of camel hooves, slowly fading into the distance.

"You see, I was right," said Solon who sometimes tended to be a know-it-all. "This was a high-level visit. Foreigners. So it is written in the stars, isn't it! The birth of this baby. I assume it's going to be busy around here. Once this is known, everyone's going to come here."

• • •

But there was no stream of visitors. It remained quiet until a few days later. Then in the middle of the night, all hell broke loose.

Stomping boots, shouted commands. "You know what has to be done. Two years old and younger. All boys. See to it that no one escapes!"

Hard knocking on doors. Rapid footsteps. Screaming. Crying.

"No, not my child! Please, spare him, please!"

"I beg you, kill me, not my son!"

"God of Israel! What are they doing! My child! My baby!"

The desperate crying and wailing were drowned out by loud voices. The door of the stall was thrown open. Men came in. Soldiers, if I'm right. They had torches with them. "Is there anything here?"

"I don't think so. No, nobody."

"Look at the tools there, in the corner. I think people have been here recently. Turn the place upside down."

The feed trough was thrown over and the straw was tossed around.

"We'll just burn the place down."

We'd heard rumors before about an opponent of the Creator. Now we saw him in action. Not personally—I don't think it's possible that we, as matter, can observe the immaterial—but evil was active here, and I could feel its presence almost physically. In the rage of the soldiers, in their determination, in the lack of any compassion. In one way or another, the opponent had somehow come to know about the rescue plan of the Creator.

I don't know what he feared from a baby, but it was clear what he wanted to achieve with this brutal action. He wanted to get rid of the Creator. This he could not do. Joseph, Mary, and the baby had just departed the previous night. According to one of our sources, an angel had warned Joseph of what was going to happen. They'd fled abruptly and would seek refuge in a foreign country where no one was waiting for them.

17

TRIAL

The house and farm buildings were burned to the ground that night, and that meant we were blown up into the wind and carried along. For weeks we floated around aimlessly. At times we would come down somewhere, on a roof, a road, or a field, then were taken upward again and wafted along. During a heavy downpour we found ourselves in the ground, somewhere in the desert near the side of a river. We ended up as part of a young tree entering through fine roots and growing upward within a strong branch. One day we saw a thin man with a beard coming. Oddly enough, I recognized him right away, even though I hadn't seen him since he was a baby. That man was Jesus. He looked tired and lonely. I didn't understand. What was he doing here, all by himself in this desolate, inhospitable place? Was he lost?

To my relief, an older man came toward him, a shepherd probably. The hospitality among nomads was known to us. I was sure this man would be helpful to him. "How are you?" he asked. "You look like you are hungry."

"I am hungry," Jesus answered.

"Then you should eat something."

I expected the man to take some bread and cheese from his knapsack and share the food with Jesus. However he pointed to some stones on the ground. "You're the Son of God, aren't you?" he said. "Just tell those stones they must turn into bread."

A stone that could turn into bread? What a strange suggestion. Didn't this man realize that bread was made up of different elements and had a totally different material structure than a mineral? That he had suggested something impossible?

But wait . . . there was yet another possibility. It could be that this man was not just a shepherd but someone who knew who Jesus really was, how powerful he was. That he, in fact, was one with the Creator. That he could turn a stone into something else without any problem.

Now that I thought about it, that seemed to me to be a good suggestion. Jesus needed refueling, I could see that. The cells in his body needed nourishment. I saw how Jesus swallowed with difficulty. How he looked at the stones that were shaped like the loaves of bread the women in Israel baked. Then he shook his head. "The holy books say, 'A person cannot live by bread alone,'" he said. "'One lives by every word that God speaks.'"

I didn't understand what those words meant, but I did understand that Jesus rejected the suggestion. And while he needed the food so badly, at that.

The stranger wasn't finished speaking. "Stand up," he said. "Do you see? We're standing on the roof of the temple in Jerusalem."

Jesus closed his eyes and opened them again. "I see it," he said.

I didn't understand. There was no temple to be seen here. Why didn't Jesus laugh at the man? Did he really think he was standing on the roof of the temple in Jerusalem?

"You're God's Son, aren't you?" said the man. "Prove it then, for all those people watching down in the square. Jump from the roof. God will save you. What's written in the holy books applies especially to you. God has given his angels the task of protecting you. You won't even stub a toe. You could use some fame, right? When you float down without getting hurt, everyone will know who you are. Everyone will know you are the Messiah. It will be as if God himself is pointing to you."

It was a ridiculous suggestion. I couldn't imagine that Jesus would take this seriously. On the other hand, what if he really believed that he was standing on the roof? If he really believed this was the way to become known and reach the people? Something in what was happening here reminded me of that event many years ago. Of Womuntu and Maisha and their tribe, many thousands of years ago, who'd followed their own wishes, against the rules of the Creator. The results had been disastrous. It seemed that Jesus was seriously considering the proposal. For one endless moment it was quiet. Then he looked at the man. "No," he said. "You may not challenge the Lord your God to prove his power."

The man was not to be dissuaded. "Oh, now look at where we are," he said, and he spread his arms wide. "See that view? From here you can see all the countries, all the mighty kingdoms on earth."

There was nothing to see apart from rocks, hills, and the barren landscape, with a bush or tree here and there. But Jesus looked around himself and nodded.

"I have a proposal for you," the man continued. "This whole world is yours. All peoples. All riches. All fame. You don't have to do anything for it. Nothing difficult. Nothing painful. Just kneel for me a moment and tell me how much you admire me, and it's all yours."

• • •

What kind of person was this? A kind of king? How could he promise Jesus something so grand? We didn't know, but the hesitation in Jesus' demeanor showed how tempting the offer was. How much effort it took for him to say no. He was clearly just a human being too.

The sun was burning. Apart from the whisper of the breeze not a sound could be heard. It was as if time stood still. As if creation were holding its breath.

Then Jesus smiled. He straightened his back. Despite his emaciated appearance, he suddenly looked like a king. "Go away, Satan," he said. "In the holy books it says, 'Kneel only for the Lord, your God, and honor only him.'"

"That's a no then," the man said carelessly. He shrugged his shoulders and sauntered away as if nothing had happened. Jesus stood still for a moment. Then he staggered and fell to his knees. The back of his tunic was drenched in sweat.

• • •

"This isn't going well," Solon whispered.

• • •

Then suddenly, out of nowhere, there was help. People—were they people?—came to help Jesus and brought him under our shadow. A woman had water and clean clothes for him; two men had baskets of food: bread, grilled lamb, grapes, goat's milk, and dates. There was prayer, eating, and laughter. Jesus recovered. He looked relaxed and relieved, like he was among friends.

• • •

"That was a little too scary for me," Ensis sighed. "I really thought he was going to do it."

"No, not me," said Aris. "I wasn't afraid for a single moment. Isn't Jesus one with the Creator? He would never break one of his laws."

"But he's also human," I said. "He's vulnerable. A person who's weakened like that is susceptible to attack. And that shepherd, Satan . . ."

"I wonder who he was," said Solon. "It seems to me he was an opponent of the Creator himself who'd hidden himself in the guise of a human being. I have the feeling that the creation has just escaped a great danger."

18

SPEECH

FROM THAT MOMENT ON we traveled with Jesus. He'd broken off the branch of the tree where we we'd been, and he used us, along with millions of other molecules, as a walking stick. We were there when he went on his way to the north of Israel and when he picked twelve men to be his students. When he healed sick people, gave back sight to the blind, and chased away evil spirits. We listened in when he told stories about his Father and about a new world that was about to dawn.

More and more men and women accompanied him as he traveled. It seemed as if he lit points of light everywhere he went. Every day it became clearer who he was. Who else but the Creator's Son could turn water into wine, calm a storm with one word, and see to it that a man with a spinal cord injury could walk again?

These were no mere tricks that he performed. No, they were miracles that changed people for real. Apparently they were a foretaste of what it was going to be like in the new world he was preparing with his Father, a world where there would be no pain and no more death. He told people about it wherever he went. We were there when he climbed a hill one day—we helped him on the steep part, I'm proud to say—and gave a speech to the people who'd gathered there about his Father's plans.

"God's new world is for all of you," he called out. "For you who are sad, God himself will be your comfort. For you who are kind. Who do God's will. Who are good to others. God sees it, even if no one else notices. God loves you as his own children. You may consider him to be your Father. And you can ask him for what you need. God is more powerful than you can imagine."

He explained to the people what his Father thought was important. Love, that was the key word. A boundless love that was much wider than what was common among people. A love that not only reached out to the Creator and to family and friends but that, in following the Creator himself, had to be there even for enemies.

Jesus corrected more misunderstandings. Apparently there were quite a few people who only adhered to the form of God's precepts without following them in their hearts. "Don't think my Father's pleased with you if you live a decent life," said Jesus. "Don't think it is enough not to murder, steal, or commit adultery. Let me put it to you straight: if you even dislike someone, if you call someone a loser, if you're jealous of their possessions, if you long for the husband or wife of another person . . . you don't belong in God's new world."

"This is too much to ask," Aris said, startled. "They really can't do that, these people."

The people listening to Jesus also became restless. "Who, then, could ever enter that world?" asked one of his friends.

"No, no one can," said Jesus, "that is, if it depended on people. But it depends on God. With him, everything is possible."

He turned to a young mother standing nearby with a baby on her arm. "Ma'am, can I hold your daughter for just a minute?" The woman laid her baby in his arms, and Jesus looked down on the little girl with a smile. "Look," he said. "If you want to enter God's new world, you must become like this baby."

An awkward silence fell on the crowd. People looked at each other. Someone cleared his throat. It was a five-year-old child who shouted, "But she can't even walk yet."

"Exactly," Jesus said. "The only thing this child can do is receive. That's how you enter God's new world. Not because of your accomplishments. Not because of your piety. You can only enter when you let yourself be carried in like a little child."

• • •

"I don't quite understand," I said. "Do they still have to obey those strict rules or not?"

"I think that's the intent," said Solon, "but the Creator also understands that no one is capable of doing that."

"How can they enter the new world then?" I asked. "I don't understand."

"That's because you're a proton," said Ensis. "I guess that's the problem."

"What I wonder about," said Aris, pensively, "is why he wastes his time speaking specifically to people who have a low position in society. Laborers. Women. Criminals. It seems to me it would be a lot more efficient if he reached out to the leaders of these people. To those theologians who are always working with scrolls. Those are the people who have influence. If he would explain to them who he is, that he's the Son of David everyone is waiting for, then they could let everyone know."

"I don't think that would work," said Solon. "He told them. He showed them. That time he healed the man with the withered hand, do you remember? There were a lot of people there. But the scholars didn't recognize him. They certainly didn't fall on their knees before him. They convinced themselves that he was breaking the rules."

"They're like white blood cells that've detected a virus and are trying to make it harmless," I said.

"It's more like cancer cells that join together to attack a healthy cell," said Aris.

"Let's not worry," Solon said. "We're talking about people here. They're not that smart. They might think they're in control, but they have no more power than the Creator gives them. They'll never be able to harm his Son."

"But that opponent then?" I objected. "That opponent of the Creator. He said he possesses the world."

"And you believe that?" answered Solon. "Haven't you seen how Jesus regularly chases evil spirits away? No, this is going to turn out okay."

19

DESPAIR

Two years passed. Jesus traveled throughout the land with his students and spoke tirelessly about his Father's plans. More and more people followed him, although there were also some who didn't want anything to do with him. Strangely enough, these were often the religious leaders.

In more and more places in Israel the light began to shine, one could say, but it was like the night sky, where despite the stars, darkness still prevailed. I waited impatiently for the moment when Jesus would seize power and make the darkness disappear.

But that didn't happen. Instead, Jesus was forcefully arrested one evening and taken to court. I was shocked. How could this be? His students, who were always with him, ran away, and his followers deserted him. I didn't understand why Jesus didn't destroy, with one word, the people who arrested him, or chase them away, and why he allowed himself to be handcuffed and taken away as if he were powerless.

During the arrest, someone had knocked Jesus' stick—our dwelling—from his hands. Another man saw us lying on the ground, picked us up, and took us to the house of one of the religious leaders.

The room was lit with oil lamps. Through the open windows, the smell of a wood fire came drifting in. I saw how Jesus stood in the middle of the room, his hands tied behind his back.

People were all shouting at the same time, accusing Jesus of the most terrible crimes. Jesus didn't say a word. He was silent, until the moment when the leader of the court asked Jesus the question, "Are you the son of God?"

Jesus replied, "Yes, I am."

Yes, I thought, *you tell them. Show them*!

Outraged cries filled the hall, and the leader of the court jumped up in anger and ripped apart his robe. Apparently that was a way of showing his shock at Jesus' response. "It is clear," he proclaimed, "that this man insults God. What is your verdict?"

There was now complete chaos. "The death penalty!" someone shouted, and all who were present approved.

"Away with him!"

"Stone him!"

• • •

"Why doesn't he do anything?" I asked, shocked. "When is he going to prove that he's actually the Son of the Creator?"

"Patience," Solon said. "It will happen. Before you know it, the tables will be turned. He can destroy them at any moment."

• • •

But that didn't happen. A couple of men who had attended the session were so angry, apparently, that they pushed to the front to spit in Jesus' face, blindfold him, punch him, and kick him.

One of them took hold of the stick we were part of and beat Jesus. Hard.

I was shocked. *We* hit him, the Son of the Creator, as if we were trying to knock him off the world he owned.

"Who did that?" said a man and laughed. "Hey, prophet, which one of us hit you?"

Jesus didn't answer.

We hit him another time, and then once again.

He did nothing.

He allowed it all to happen.

We landed on the ground, were kicked outside, and didn't know what was happening during the next few hours. One of the friends of Jesus—we knew his name to be Peter—found us, picked us up, and took us along to a hill outside the city. He didn't pay attention to the other people walking toward the hill; he paid no attention to where he was walking and just kept weeping, "Forgive me. Forgive me. Oh Lord, forgive me. What have I done . . . ?"

From a distance we saw what was happening. It was Jesus. They'd hung him on a pole, and his arms were nailed to a crossbeam, as if he were a serious criminal. His body was bruised and battered. He was bleeding and sweating, and his breathing was labored. No one came to his aid. The people who were watching even made jokes about him.

"Now look at him. Isn't that the man who said he was the Savior of Israel?"

"Hey, Jesus! Show us how powerful you really are! Come down from that cross!"

"Yeah, come down from that cross. Then we'll believe in you!"

I was stunned. Why didn't he do it? Why didn't he pull himself loose (as, long ago, one of the judges of these people had done after he'd been tied up)? Why didn't he destroy them, these people made of stardust? Why didn't he show them who he really was?

He didn't do anything. He just hung there, his eyes closed, his face contorted in pain. Powerless.

Two men hung beside him, one on either side. The first pulled himself up a bit, cursed because of the pain, and said, "Hey, you there. If you're the Messiah, this is your chance, right? And you can save us too, at the same time . . ."

"Shut up!" the other man snapped at him. His anger was clear in his voice "We're going to die, man. Let that sink into that stupid brain of yours. Soon we'll stand before God. Aren't you afraid? I am. There's a reason we're hanging here. I know what I've done. But this is Jesus, you know, the only one to stand up for people like us. He's never done anything wrong, the way I see it. Jesus shouldn't be hanging here." He paused, took a shortened breath, and said, "Jesus, will you think of me when you're king, soon, in your new kingdom?" Jesus looked to the side. He nodded slightly, and said, "Listen well. Even today you will be with me in heaven, in a place that is more beautiful than you can imagine."

A hot wind began to blow. Sand and dust darkened the sun. It was in the middle of the day, but the light disappeared. It was as if the Creator himself withdrew his hands from his Son.

It went quiet. Although it wasn't completely dark, it did remind me of the time we'd spent in real darkness, now long ago, long before the first star was formed. Light was chased from the earth, and the darkness seemed endless—not only for us but also for the man who hung there dying, utterly alone.

Many of the spectators went home, but a few people stayed behind. Because they had a task to complete. Because they wanted to know for sure that Jesus would really die and not be freed at the last moment. Or because they didn't want to desert him.

I recognized one of the women as Mary, the mother of Jesus. She stood close to the pole where Jesus hung, and looked up to him, tears streaming down her face. A man stood next to her—John, one of Jesus' students who'd become a close friend over the years.

Jesus pushed himself up a bit, moaned for a moment, and cleared his throat. "Mama," he said. His lips were cracked, his were eyes bloodshot, and his voice was hoarse. "He is now your son." He closed his eyes, opened them again, and then turned to John. "She is now your mother."

The wind died down, the dust had settled, and the sun reappeared—weakly at first, but then it shone more and more brightly. Jesus opened his mouth. It was as if he was trying to say something with his last bit of strength. One of the soldiers jumped up. He cast a glance at the three men who hung above him as if he wanted to make sure they wouldn't escape, then walked to a wine vat, dipped a sponge in it, pushed it onto a stick, and walked back to the place where Jesus was hanging. He lifted the sponge to Jesus' mouth and wet his lips.

Then Jesus pushed himself up one last time and looked around with the look he'd had in his eyes when he'd calmed the storm, and when he'd chased away evil spirits. "My work is finished," he cried out. It sounded like he had won.

Then he died.

• • •

The Son of the Creator was murdered by the people he'd wanted to rescue, the people he loved so much. The great opponent had succeeded in his intent. From the moment that Jesus had been born, the opponent had tried to defeat him, doing everything he could to sabotage the rescue plan. He had succeeded.

The sun shone again, but it was as if darkness lay over the land as a suffocating blanket. We felt a deep subterranean rumble. The earth trembled for seconds. Jesus' friends hardly seemed to notice it, but the Roman soldiers were visibly upset. They fell completely out of their role as heroic guards of order. "An earthquake!" "Let's get out of here. Out of here. Away from that cross. It's going to fall over!" "Oh, Jupiter, save us!"

When the ground stopped quaking, they looked at each other. "That was really bizarre, man."

"Exactly at the moment when Jesus died!"

"I thought I'd had it."

"No doubt about it," their leader said. "This man was the Son of God."

20

SHOCK

"Didn't I say so?" I shouted when Peter had stood us up against a wall. "I told you things would go wrong! And you didn't believe it. You said things were going to turn out well, Solon. But that didn't happen, did it?"

"Maybe you're right. I just don't understand it," said Solon. "I don't get any of it. I thought he was getting more and more supporters. I thought he could be proclaimed king at any moment. Even last week, when he traveled to Jerusalem and rode on that donkey. Now it's beginning, I thought."

"Yes, you thought that," I said. "But now he lies buried in the ground."

"Technically not in the ground," Ensis began, "but in a . . ."

"How could things go so wrong?" I asked. "The Creator has been working on it for so long. He's been busy with it for millennia, preparing for the rescue."

"Maybe he didn't take into account that the opponent had become so strong," said Solon.

"I think he made a big mistake when he went to Earth himself," said Aris. "He could've sent one of his personal envoys. That Gabriel, for example."

"Maybe he made a mistake working with humans," I said. "Maybe people are much, much worse than he thought. I hardly dare to say it, but I fear that things have gotten out of hand for him."

• • •

It was the third day since Jesus had died, and we found ourselves in a home where some former supporters had gathered—family, friends, his students, men and women who'd always supported him, that is until the moment that he was arrested. So Peter was there, and John. Like the rest of the people there, they still seemed to be in shock.

From our place against the wall, my friends and I saw it all with mixed feelings.

"Now what?" Ensis asked.

I thought about it. Honestly, I had no idea. The Creator had given everything to save the people, and they'd clearly indicated that he was not welcome in his own world.

Would he punish them? Allow them to experience what had happened to his Son? Or would he turn his back on them and not show himself again, considering the "human project" to be a failure?

Eventually people would become extinct, and other animals would replace them. Much later, our star would become old and red, then die, and we would start again, somewhere else in the universe.

For the first time in my life, I felt a sense of emptiness. Of futility. As if I'd been allowed to see my destiny for a moment and then was told that I'd never reach it.

Suddenly a young woman came running into the house. She was out of breath and seemed distressed. "He's alive," she said breathlessly as she took hold of a table. Her long hair was in tangles. "We have seen him!"

"What?" said John. "What're you saying?"

Two other women came inside. "He's alive!" one of them shouted. She tripped over a jug that was in her way and barely managed to keep herself from falling. "Jesus is alive!"

John shook his head. "It's not going well, is it? We all feel that way. And what's happened is truly terrible. Here, sit down for a moment, and let me get you a glass of water."

"No!" the first woman shouted. "You have to listen. We have been to the grave and . . ."

"Didn't we say there was no sense in going to the grave?" said Peter with irritation in his voice. "There are soldiers on guard. They're not going to let you come close."

"It was empty! The grave was empty. The stone was rolled away. There was an angel who said that he'd risen from the dead."

It became quiet in the room. Someone cleared their throat.

"Could it be that you dreamed it?" another man asked. "I mean, we're all having a hard time, and we're not getting much sleep. That's happened to me too. Sometimes I suddenly think I see him, outside." He coughed and wiped his eyes. "Or that I hear his voice."

"Oh, I get it," the woman said, and looked at him with disappointment. "You think that we've imagined it. Well, this isn't a fantasy. I have seen him with my own eyes. Oh, yes, I also need to pass something on to you, Peter. You should all go to Galilee. He will meet you there. He said you would know where. You have an appointment, he said."

Peter and John looked at each other. And all of a sudden, Peter grabbed the stick that we were part of and ran out the door with us, into the city streets. He was followed by John, who soon caught up to him.

• • •

"What's happening now?" asked Aris, while we flew though narrow streets.

"I can tell you that," I said. "We're going to that cemetery, and then we'll see it's nonsense. We'll either see a closed grave or an open grave with a decomposing body."

• • •

"And?" gasped Peter when we arrived at the grave.

John shook his head. "Look for yourself. His body is still lying there." Together with Peter we entered the cave in the rock. It was quite dark and, at first sight, there seemed to be a body lying there. As we got closer, however, we saw something strange. Cloths were lying there, but there was no body.

Peter staggered out of the cave again, banging his head against a sharp rock. Blood dripped from a wound in his forehead, but he didn't seem to notice. "They were right," he whispered, and he let himself drop to his knees in the grass.

"What're you saying?" John also went inside. When he returned he was pale. He sat down beside Peter. "This is the grave . . . right?"

Peter nodded. "There's no doubt about it, I'm absolutely sure. They buried him here."

"And he is no longer here. That can only mean one thing . . ."

Jesus' two friends were now firmly convinced that Jesus had come back to life. He had risen from the dead.

• • •

"I'm sorry," I said while we returned to the city, "but those people are really gullible. That grave is empty, I can see that. That's remarkable, at least if it's true that soldiers were guarding the grave and a boulder was rolled in front of it. And about those cloths, that's also strange, but it doesn't prove that he came back to life. The body might've been taken away by someone. That's even more likely."

"Exactly," said Solon, who agreed with me for once. "Of course, people have a great ability to imagine things."

• • •

But that evening he suddenly stepped into the house, and we saw him ourselves. Jesus. The son of Mary. The Son of the Creator. No phantom. No ghost. A real person, warm and alive, his cells with the same DNA that he'd had before his death, with a heart muscle that beat, with blood vessels through which blood flowed, and with eyes that had been sightless but now could see again. His wounds were healed, and his body somehow seemed more powerful and radiant than before. The scars in his hands and feet, however, were still clearly visible. He hugged his friends, and he ate and drank with them.

"Ah," he said. "My stick. You saved it for me. Thank you. But Peter, you keep it, because I don't need it anymore. I have to leave, but I'll see you all soon in Galilee."

Then he left.

• • •

Never in my almost fourteen billion years had I felt such joy. I couldn't for a minute comprehend what'd happened, but I understood that it was something that turned the world completely upside down.

"This is totally impossible," said Aris.

"If I hadn't seen it myself, I wouldn't have believed it," said Solon.

• • •

A bit later, one of Jesus' other friends came in, Thomas. He hadn't been there earlier that evening when Jesus had stopped by, and he didn't even consider believing that it had really happened.

"I'm really sorry," he said. "I'm happy for you that you've seen him. Jesus continues to live in our thoughts and dreams. That is beautiful. But don't tell me now that he was really here, because that is just impossible."

"It's true," Peter exclaimed. "I swear it, he was standing here. Right where you are standing now."

Thomas shook his head. "You're just making things more difficult for yourselves. What has happened is terrible. And it's terrible that we have to move on without him, but let's not tell ourselves that he's still here. That's no help to anybody. Not to her either." He nodded to Mary, the mother of Jesus, who was also present.

"He's alive," said Mary. She came to him, beaming, and put her hand on his arm. "You have to believe it, Thomas."

Thomas sighed. He looked at Jesus' mother as if he felt sorry for her, but his words were harsh. "I'm sorry. That just doesn't do it for me. Before I can believe something like that, I have to have some real proof. If you want something very badly, you can just start imagining that it really took place. You're not the first, and you won't be the last." He looked defiantly around the room. "As far as I'm concerned, you can believe in this fairy tale if it gives you comfort. I'd rather continue to use my intelligence. See first, then believe. That's what I learned as a child from my father, and I'm going to stand by that."

"But haven't you seen the things that Jesus has done?" said Peter indignantly. "How he just stilled the storm. How he changed water into wine. How he healed people."

"You saw the bad shape I was in," said Mary, not the mother of Jesus but another woman with the same name. "Evil had me in its power. My life was hell until he chased away the spirits with one word."

"I know all that," said Thomas. There was despair in his voice and tears in his eyes. This clearly meant more to him that I'd thought. "I don't want to take anything away from that. I loved Jesus as much as any of you did. I saw his power with my own eyes during the time we traveled together. But I also saw that he died. He is dead. All that's left of him is a body, a corpse that's now decaying." He raised his hands apologetically to the mother of Jesus. "I'm sorry that I have to use such harsh words, Mary. But it is what it is. I'm not going to believe the story that he has come back to life. I'm not going to be carried along by your fantasy. I won't believe it until I see it with my own eyes and touch him myself. But that isn't going to happen."

No matter what the others said, they couldn't change Thomas's mind. He continued to consider the story a delusion.

And then, a week later, Jesus was standing in the room again. Just like that.

• • •

"Look," I said, happily. "Look who's here!"

"Where did he come from?" asked Ensis.

"No idea," said Solon. "He is the Son of the Creator. He's not bound by our laws."

• • •

The men and women in the room saw him too. They jumped up and ran toward him. However, Jesus only seemed to have eyes for the man who couldn't believe that Jesus had returned from the dead.

"Thomas," he said, and he reached out to him and hugged him.

Thomas became so pale that we thought he was going to faint.

Jesus patted him on the shoulder and laughed. "Don't be afraid, I'm really not a product of your imagination. Come on, sit next to me. You wanted to see my scars, didn't you? Look." Jesus reached out, and Thomas hesitated. But then he carefully stroked the scar on Jesus' wrist with his rough index finger, a scar that was already fading. He swallowed. Then he sank to his knees. "My Lord," he stammered. "My Lord. Now I know. Now I know that it is you."

"You believe in me because you have seen me," said Jesus, "because you have touched me. You didn't want believe it from your friends' word."

Thomas nodded, his head bowed down.

"Don't feel guilty, Thomas. Because you couldn't believe the message about my resurrection to be true, you're going to be able to help many people. From now on, many people are going to believe in me without actually seeing me."

21

REVERSAL

"I BELIEVE," I SAID, "that we have underestimated the Creator a lot."

"Speak for yourself," Ensis said indignantly. "I would never doubt the Creator."

"Pro is right," Solon said. "We all thought it was over, the rescue plan. We all thought that after all these thousands of years of preparation, it had still ended in failure. Now we discover that's not the case. Such a tremendous surprise can only come from the Creator. Imagine. For four billion years we have known only that death is the end, that it is an irreversible process."

"Then what about that friend of his?" I asked. "Remember Lazarus?"

It was still clear in my mind. A few weeks before his arrest, Jesus had been called to see a sick friend. We went there, but when we finally arrived, it was already too late. A few days before, Lazarus had died and had been buried. Jesus had brought his friend back to life, which had been a huge shock for all the people who were there and, of course, also for us protons and neutrons.

"Yes, exactly!" said Solon. "Lazarus. That really blew our minds, forgive my metaphor. I now think it was a foretaste of the great miracle that he had planned, of what Jesus was going to do, of how he was going to conquer death."

"Hold on a minute," Ensis said. "Do you mean that it was all intended? The arrest? The murder?"

"That's just impossible," I cried angrily. "What happened to him was the greatest crime that we've ever witnessed. It went against everything the Creator ever intended. How can you say that it was intended, Solon?"

"Take it easy, Pro," said Solon. "There's no way that the murder can be condoned, but the Creator is the Almighty, the origin of all that's good. Evil can never beat him. Even if it strikes with all its might, he will eventually turn it into something good, just as he let the dark be overcome by the light. The night can never be the winner."

• • •

Solon was right, even if we didn't discover this until much later. If I understand it correctly, the death of Jesus had ensured that human beings no longer had to do anything to come into favor with the Father. Everything they had done wrong—whatever it was—the Creator didn't hold them to account but forgave them. He even regarded those human beings as if they were Jesus himself and had done nothing wrong. As if they were the people he'd always had in mind, the people he'd chosen. It was a complete reversal.

We had no idea what was going to happen now. Would a peaceful society arise spontaneously, in which everyone would know the Creator? Would no one die from that moment on? Would Jesus perhaps become the king of Israel to ensure that everyone would live as the Creator had intended? We weren't the only ones with questions. Jesus' friends didn't understand what the Creator's intention was either. They were ready to take up weapons to expel the Roman occupation, but that wasn't the Creator's intention. Jesus made clear to them that he had something completely different in mind. He had to go back to his Father, he said, to arrange things for the time when he would come back to stay. In the meantime, his friends had to go out to make the good news known everywhere and to invite people to make the Creator part of their lives. Everyone who wanted to belong to him would let themselves be baptized in order to make a new start, a life such as the Creator had offered to the first people, a life with Jesus that would last forever.

His new followers weren't to confine themselves to the people who descended from Abraham. No, the invitation was meant for all people. Whether Jesus' friends understood all of that, we didn't know, but it was great news for us. News that, I must admit, we nevertheless received with some skepticism.

"All people on earth?" I asked. "All people? Does he know how many people there are now?"

"Three hundred million!" Aris called out. "Three hundred million! That's what I've heard."

"That's what I mean, and they have to reach all of them? Impossible! There's just too few of them."

"That's right," Ensis said. "Assume that they had enough people. It still would not be possible to reach all of them. The continents are now separated by distances that are just too great."

Ensis was right. I remember the time, long ago, when there was one large landmass surrounded by the great sea. But the continents had gradually drifted apart, and the pieces of the puzzle had been pushed ever farther apart.

"Exactly," responded Aris. "There are continents where they will never be able to go. Besides, how will they be able to communicate with people from other continents, with all those other languages that they speak?"

"Just wait and see," said Solon. "This is the Creator's Son, right? When he has a plan, it's doable. For him the earth is just a speck of dust. For him the oceans are just a drop. And for those languages, I assume he'll find a solution. Don't you hear what he's telling them, that he will stay with them until the new earth comes?"

• • •

During a conversation between Jesus and Peter, we found ourselves on a hill close to Jerusalem. We hadn't yet seen the sun that day. It was chilly, the sky was gray, and the clouds hung low. Suddenly a fog closed in, a fog so dense we could hardly see. It didn't last long. The fog lifted, the clouds parted, and the sun broke through.

"What?" I heard Peter say. "Where are you, Lord?"

"Where did Jesus go?" I heard John ask. He looked around. He looked upward—which was strange because, of course, people can't fly. "Did he go up?" asked one of Jesus' other friends. There was a note of bewilderment in his voice.

"I don't understand it," said John. "It looks like he has dissolved into that cloud."

"Or that he was taken along by that cloud," said a man called Andrew.

"That cannot be," said someone else.

Thomas scratched his head. "I never thought I would say this," he said slowly, "but could it be that he has been taken up into heaven?"

"Right," a voice rang out. Just next to us stood two people that we had never seen before. They told Jesus' friends that he had gone over to another world. "This doesn't mean that he is gone forever. He will return one day, in the same way that he departed."

It struck me that Jesus' friends weren't sad or overwhelmed by this sudden departure. It seemed as if they had more or less seen it coming. They said a prayer of thanksgiving and hugged each other as if it were day of celebration. Then we went back to Jerusalem with them, where they told the news to their friends and the mother of Jesus.

There was a festive, expectant atmosphere in the house where we stayed for the next few days. Peter kept saying how much he would like to go back to work, to tell all the people in Jerusalem and beyond about Jesus. He couldn't really wait, he said. This was such good news.

"Just be patient a little longer," John said then. "We have to wait. That's what he said."

"I know," said Peter.

The friends were apparently waiting for a person called Breath, who Jesus had often talked about. Breath was going to help them with the task awaiting them. We didn't know this person, but it was someone who seemed to be closely connected to Jesus and his Father. I wondered if it might be a high-ranking angel.

During the days that followed, we frequently talked about the curious moment when Jesus had disappeared, and the fact that his friends had been staring at the sky. "Where do they actually think the Creator's residence is?" I wondered aloud.

"They were talking about heaven," Ensis said. "So if you ask me, they think he lives up there somewhere, between the moon and the earth, perhaps, or between the stars."

I laughed. "Doesn't it occur to them that there might be another dimension," I asked, "one that the Creator can step in and out of? Where did they think Jesus was in the last few weeks, in the days he didn't spend with them?"

"I don't think they can comprehend such a thing," said Solon. "I think they know only three dimensions, or four if you include time."

"People are different from us, Pro," said Solon. "They have such short lives. There's so much that they don't understand."

"Well, Solon," said Aris. "I'm sure there's something you don't understand."

"That could be," Solon admitted. "I still don't understand much about love."

"And that while we've been together for such a long time," said Ensis. "I am deeply disappointed in you, Solon."

22

FORWARD

BECAUSE WE WERE PART OF THE WALKING STICK, we were able to follow Peter wherever he went. We were there when, a little over a week later, Breath arrived—the helper Jesus had promised. We didn't see him, just as we can't see the wind but can see the trees sway in strong winds. But we noticed he was present by the remarkable things that happened on that day—by the fact that Peter, an unschooled fisherman, suddenly ran out into to the street and began to tell people from Crete about Jesus in fluent Greek, for example. And by the fact that his friends did exactly the same—they spoke about Jesus in languages they'd never learned, to my knowledge. And the story they told hit home like a bomb. The people in Jerusalem who listened to them—the same people who'd demanded Jesus' death two months earlier—were now asking themselves what they'd done. This had to be the work of Breath. Thousands of people joined Peter and his friends that day and let themselves be immersed in water, a ritual that made it clear that they wanted to wash away their old existence and start a new life.

Within a few days, Jesus—who himself was no longer present as far as we could tell—had more followers than he'd had in all the days he traveled throughout Israel. It was not a whim, not temporary hype, as the authorities hoped. The new followers of Jesus began to form a community that was a bit like a large family, a family like the one the Creator had envisioned long ago. They called each other brothers and sisters, and that's how they got along with each other too. They were interested in each other, helped each other, shared their belongings with those who were in need, and cared for people who were ill or old.

Every newcomer was welcomed with love and greeted like an old friend, regardless of their gender, rank, or social position. Rich and poor, landowners and slaves, men, women, and children, they all came together to eat, to sing, to talk about Jesus, and to pray. I especially found the latter an interesting phenomenon. These people spoke to Jesus as if he were sitting next to them in the room, even though they didn't see him or get an answer from him, as far as I could tell. Apparently, he was very near to them and was able to hear and help them.

The community spread rapidly via Jerusalem to the rest of the country, and then across the border. Breath was invisible but, at the same time, I think he was everywhere. And somehow he brought Jesus close to the community. The lights went on in more and more places in Israel and beyond. From the beginning of the new period, we also heard of things that went wrong. It seemed that the opponent of the Creator had started a desperate offensive, right in the places where the light shone. Jesus' followers were opposed by the authorities, imprisoned, and mistreated. Some were even killed, as if they were notorious criminals. There were also people who'd come to know the light and had allowed the dark back in or sometimes even tried to extinguish the light.

I found the latter particularly incomprehensible. "I don't get it," I said. "They were so happy when they heard about Jesus. They've come forward to be baptized. They've become members of the Creator's family. How can they possibly go back into the night?"

"Exactly!" cried Aris. "Why doesn't the Creator stop them?"

"Or why doesn't Jesus stop them?" I said.

"Would you like it better if he would hold them captive," asked Solon, "if he were to prevent them from running away from him?"

"Yes," I responded. "That'd be much better for them."

"That's not the way the Creator works," said Solon. "He has so much more respect for people. He'll never force them or hold anyone against their will."

• • •

Together with Peter we traveled throughout the whole country. It was as though we saw Jesus himself at work. Wherever his name was mentioned, sick people recovered, paralyzed people could use their legs again, and sometimes even death disappeared.

Peter and his friends were in constant contact with Jesus. They didn't see him, but they spoke to him like he was sitting next to them, as if he was still with them. Sometimes he seemed to answer them, or he spoke to them in a dream—to clarify things they didn't understand yet, for example.

One day Peter took us into the house of a Roman officer, and he greeted the man like an old friend. That was odd because the officer wasn't part of Abraham's family. According to the sacred laws, Peter wasn't allowed to visit him; he wasn't even allowed to touch the man.

But Peter described that he had heard in a dream that those rules had been lifted, that the family of Abraham was being expanded, and everyone who believed in Jesus should be allowed to belong. This also had consequences in all kinds of practical areas—the new followers of Jesus didn't have to follow Israel's food laws from the old holy books. We noticed that this was an earth-shattering discovery for Peter and his friends. The Creator had not only Abraham's family in mind, but he wanted to rescue *all* people worldwide. He wanted to embrace the entire sighing, toiling creation in his plan of deliverance.

One day, Peter was arrested rather forcefully (he just wouldn't stop talking about Jesus). His stick, our dwelling, was taken from him, broken into pieces, and thrown into the fire. We never saw Peter again.

• • •

How did things end up with us? The usual story. We floated through the air, met many new protons and neutrons in neighboring atoms, and ended up in the water a few days later. We connected with other molecules, became part of a sea snail, were eaten by a lobster, which was captured by a large puffer fish after an exciting hunt. Eventually we ended up somewhere in shallow water, became part of a papyrus plant, were harvested and, together with countless others, processed into a roll of paper. Now that was something entirely different! One day when we were being unrolled, we saw a familiar face.

"Hey, isn't that one of Jesus' friends?" Aris asked.

"Yes, it's John!" I said. "But he's so old now!"

"That's what happens with people, Pro," said Solon. "They're not like you and me. They wear out. It's the cells, right? The DNA becomes more and more damaged. At some point it can't be repaired any longer."

"I know that," I said somewhat offended. "I know how cells work, Solon."

"But what is that man doing here," Ensis interrupted, "so far away from his homeland?"

"Don't you remember what Jesus said," answered Solon, "that they had to go throughout the world to make the good news known?"

"If that was his plan, he could've chosen a better place, where more people live," said Ensis. "It seems to me that there's not a lot of opportunity here to make Jesus known."

We found ourselves on a small island with very few inhabitants. Gradually we discovered why John was here. The emperor had banished him to the island to stop the rapid spread of the "Jesus movement."

John was not to be dissuaded. He had enough to keep himself busy. A friendly sea captain named Gaius, who brought guards to and from the island and regularly delivered food, brought John papyrus and ink so he could write letters to the followers of Jesus who needed encouragement or information. It was a pity that we couldn't read what was being written on us, but occasionally we heard John talking about it to Gaius.

"I don't have much time," said the captain. "The weather is going to turn this afternoon, so I want to leave in an hour. I just came to pick up those letters that you wanted to send along with me. Sorry, John, I wish I could do more for you."

"You do a lot for me," said John. "It's fantastic that you brought me some more papyrus; that will keep me going for a while. When are you coming again?"

"In two months. Can I bring you something special, something you like to eat?"

"I have everything that I need here. But papyrus and ink are always welcome."

"You keep writing, don't you?"

"As long as Jesus expects me to write, I will write. In the last few days he's shown and told me so many things. I can't write them all down. The world would be too small for so many books."

The captain sat down beside John. We saw him from the table where we rested. "You're making me very curious. What did he show you?"

"I'm going to let you read that, Gaius, the next time you come."

"Reading isn't my strong suit, remember? Maybe you can just tell me briefly."

John laughed. "Well, I'll just tell you that the Lord is working hard for us."

● ● ●

"The Lord?" asked Ensis.

"That's what he calls Jesus," I said. "Quiet. I want to hear what he's saying."

● ● ●

"The Lord has informed me that there will come a time when evil will no longer have a place on earth. I saw it with my own eyes, Gaius. In a vision I saw how death and the kingdom of death will disappear. I saw how God made everything new, how he built a beautiful capital city for his new world. How he came to live among his people, and how he comforted everyone. I saw a world without pain, without war, and without sorrow."

John smiled. "The light I saw, Gaius, was the most beautiful light I've ever seen. It didn't come from the sun or moon, it came from God himself." The old man looked at a butterfly that fluttered past the entrance of the cave. Far away the sunlight shone on the waves, and the wind rustled through the leaves of the trees. "And there was a well," he continued. "A well with water that gave eternal life. Everyone could drink from it."

Gaius cleared his throat. "That sounds so nice. Almost too nice to be true, to be honest. I'm sorry to say this, but are you sure you didn't just dream it? I mean, you're here all by yourself; I can imagine that could do strange things to a person. I'm sorry, I don't want to offend you."

John laughed, "Yes, I'm very sure. I didn't make it up. The Lord has shown me that this is going to happen, and he wants me to write it all down because he wants all his followers to know about it. You know, Gaius, it's true. It agrees exactly with what Jesus told us when he was still with us. He said he would return one day, but until now I had no idea how I was to understand that. Now that I saw some of it, I can't wait for it to happen." Joy radiated from his face. "Everything is going to be different. Everything is going to become as God intended it to be."

Gaius frowned. "But the emperor . . ." he began.

"The emperor now seems invincible, I know. The Roman Empire is so large and powerful; it seems like it will always exist. But the Lord has shown me who is really on the throne—his Father. So a day will come when even the emperor will bow to him, a day when evil will be completely overcome. Be sure to tell that to the congregation that meets at your house, Gaius. They have to know. This is a message from the Lord himself."

A small boy on bare feet came running from the wharf.

"Captain, they want to know when you're coming."

"Tell them I'm coming right away." Gaius stood up, took the letters that were lying there, and hugged the old man. "I have to go now, brother. May the Lord be with you. I will see you in two months."

He ran toward the water, stopped, turned around, and shouted, "Or earlier if the Lord returns."

23

SPACE

We found ourselves in a small galley in a space station, some 250 miles above the earth. Below us stood five people around a table. They were trying to put toppings on pizzas while the crusts were trying to escape, which was a funny sight for us too.

"Isn't it clear?" said a woman who was called Jada. She had black dreadlocks, which she'd pulled into a ponytail at the back of her head. Like the others, she'd hooked her feet under a rod on the floor so she wouldn't float away. "That view is just unbelievable."

"A couple of years ago it was even more beautiful," said a bald man in a light blue polo shirt. He grabbed his floating pizza out of the air and began to carefully top it with tomato sauce. "During that Covid pandemic, when everything was closed down. Unbelievable how quickly the smog disappeared and the earth began to recover. Although, of course, I'm glad they found a vaccine and that people don't die from it anymore."

"I hope you're right," said another woman. "My father is in the hospital with pneumonia. It could be yet another new virus strain. I'm praying to God that he'll survive."

"Oh, that's so hard," said Jada.

"I don't mean to annoy you, SaeJin," said the man in the polo shirt. "I'm sorry that your father is in the hospital, and I hope he'll recover soon. But I don't believe there's a God who's going to do anything about it when you pray to him. I really don't get it. You've finished two university degrees. You've won so many prizes in your field. You're plenty smart. How can you still believe in God?"

Ah, this was interesting. I looked at the woman that was being called SaeJin. Her shoulder-length black hair was pointing in all directions, thanks to the lack of gravity. She took a pair of scissors and cut open a small plastic bag. "I don't see the problem at all, Gary," she said as she began to put slices of mushroom on her pizza. "I really do believe that God exists. And I am a dedicated fan of science and technology. I just don't see how these two things are in conflict with each other. My faith and science go together beautifully. So when my father is sick I send him to the hospital and pray to God to heal him. Eh . . . anyone still want some mushrooms? Before I take them all myself?"

Jada reached out, "Yes, I'll take some."

"Pepperoni?" said a young man while he opened another small bag. "I'm going to put some pepperoni on mine."

"Seriously, Bob?" His neighbor rolled his eyes. "I thought you were a vegetarian."

"I know, I know. Tomorrow I'm vegetarian again, I promise, but today I cannot resist having some pepperoni."

The bald man in the polo shirt, Gary, slid his pizza into the small oven under the desktop and placed his hands on top of the desk. "You know what I think?" he said while he looked at SaeJin. "I think that God is a figment of the imagination of people from thousands of years ago, people who couldn't understand reality and therefore conceived a higher power. Where does lightning come from? How did the world come into being? What happens when there is a solar eclipse? If you have no idea, you think up your own solutions. Surely we now know how the universe came into being. We know how evolution works. Now that we have science, we've surely outgrown these gods, haven't we?"

• • •

"What?" I cried in disbelief.

"Shhh," said Solon. "Don't always talk when I'm trying to listen, Pro."

• • •

SaeJin shook her head. "Such a god to fill the gaps, I just don't believe in that, in a god that's an explanation for things we don't yet understand. I believe that God has the whole world in his hands and that he is also in control of the things we do understand."

"Exactly," said Gary. "And that's where you're wrong, SaeJin. You believe it."

SaeJin did not look worried or irritated. She took a cloth, wiped her hands, and set a timer. "Much of what I know points to that," she said, "and also the things we know from science. The laws of nature, for example. They fit in perfectly with the existence of a lawgiver. Or the Big Bang. You have no idea what caused it. I believe it's an act of creation by God. And then there's the fact that we've discovered that all the physical constants in nature were precisely tuned, just after the Big Bang, so that life could come into being billions of years later. It's simply not reasonable to ascribe all that to pure chance. Surely such things point to a God who wanted this world."

"Yep, we've had this discussion before, haven't we?" said one of the other men. "As I said back then, Who says that this is the only universe? It could well be that there's an infinite number of parallel universes. And with an infinite number of possibilities, it's not surprising that life began somewhere.

Besides, there's a good chance that there's life somewhere else in the universe."

"The problem with your reasoning is that there's not a single bit of evidence to support the existence of such a multiverse. It is pure speculation."

"Just like that God of yours," Gary responded.

SaeJin smiled. "Do you know what Pascal once said? 'The heart has its reasons that reason does not know.' It is those reasons of the heart that make me believe in God. I experience a deep sense of meaning, of the meaning of life. I don't think we're just here for no reason at all. It all serves a purpose, this universe, the earth, life, everything."

She glanced quickly through the window where the world at night was visible—a dark sphere dotted with points of light, with, over it all, the greenish glow of the atmosphere.

"I believe it has a purpose. I believe in the dignity of every person. I believe that some things are absolutely, morally wrong. I experience the existence of God in my life. And all those things I would have to think of as illusions if I were to become an atheist, such as: *yes, you may think all that, but it's actually your brain that's fooling you*. Well, I'm sorry, I just don't believe that. I don't have enough faith to be an atheist. Whereas, if you believe in God . . ."

She was interrupted by the oven timer.

"The pizzas are ready!" said Jada. She pulled herself toward the oven, opened the little door and pulled out the pizzas. Gary took off the aluminum foil and flung the pizzas like Frisbees to the others. "Does someone want part of my salami pizza?" Bob asked. He let the pizza spin on his finger as if it were a basketball. "You, Sergei?"

"No thanks, I'm going to stick to my own."

"I'd like to try it," said SaeJin, "then you get half of mine."

Bob took a pair of scissors and cut his pizza into two. "If you take the olives off first, because I don't want those."

"But just think about it, SaeJin," said Gary after he had finished the last part of his pizza. "If it's true, why don't you hear anything anymore from that God of yours? Wasn't that Jesus going to come back? It's now more than two thousand years later, and nothing's happened yet."

SaeJin smiled, wiped the scissors clean with a piece of paper and was going to answer. At that moment Jada interrupted the discussion. "This discussion is all fine, but we really do need to get back to work. Those repairs have to be finished today."

"Wait," said Gary. He pulled a silver little pouch loose from the Velcro attaching it to the table, and grabbed a spoon. "We have a little dessert. Let's quickly take care of this chocolate cake before we go back to work."

• • •

"Did he say two thousand years?" chuckled Ensis when they were gone. "Did he really say that he couldn't believe in the Creator because two thousand years have passed and Jesus hasn't come back? I had to laugh about it too. What was two thousand years compared to the billions of years that passed since the moment that the Creator began with his project?"

"And they're so convinced there can be no Creator," I said. "That's something new, isn't it?"

"Something from the last couple of hundred years," said Solon. "Quite a bit has happened lately, anyway."

"Still, it's remarkable," said Aris, "that they have discovered us, and they don't realize they're looking at the work of the Creator."

"Quite remarkable," I said, "especially for creatures that're so highly developed."

I thought back to the moment when I'd heard we'd been discovered more than a hundred years ago. The story of Ernest

Rutherford had come to me via the grapevine. I'd found it particularly amusing that we, protons, had been discovered before our good friends, the neutrons. But I didn't want to brag about it (partly because I didn't want to be reminded that electrons were discovered before we were).

• • •

"Nervous?" Gary asked an hour later. We'd been sucked from the ceiling by air currents, and now we were located in a part of the space station where we hadn't been before, on the gloves of SaeJin. "Your first spacewalk."

"Not really," said SaeJin while she was inspecting her outfit one last time. Just like Gary, she wore a voluminous white suit to protect her against radiation, supply her with oxygen, and keep her body at the right temperature. She spoke, just as he did, via a microphone in her helmet. "I'm not nervous."

"Because we all were, the first time."

"I'm not nervous, Gary."

"You know what Yuri Gagarin said during his first trip in space? He said that he didn't see God there."

"Small correction. He didn't say that himself. It was a statement by Nikita Khruschev, the man who tried to eradicate Christianity in the Soviet Union."

The hatch slid open and revealed the deep black of the universe. SaeJin grasped the handles and left the space capsule. She took us along with her. From her glove I looked at the earth that had been our home for such a long time, a bright blue ball, half lit by the sun. We saw the oceans, the continents where we'd lived, the white icy plains of Antarctica, the rusty brown of the deserts, and the great rivers that twisted across the land like dark ribbons. We saw the clouds that cast dark shadows on the water, the mountain ranges, and the green of the rainforests. For the first time in billions of years we were back in space.

For at least four hours, SaeJin was busy with her colleague, fixing a fault in the electrical supply. When she went back inside, we were left outside, stuck to the outer wall of the space station.

• • •

"Hey," said Ensis. "What's happening now?"

"We're home again!" replied Solon. "Back home in space."

"But that's not what I want at all," Ensis protested. "I want to go back to Earth! I want to see how it ends."

"I don't think that it makes much difference what we want," I said.

Not much later we collided with a tiny piece of space debris. We didn't stand a chance against the object that wasn't even a millimeter in diameter. With a bang we were launched into space, moving farther and farther away from Earth.

Soon the space station was out of our sights. The Earth became smaller and smaller, the Milky Way was visible against the dark background of the dark, and the millions of stars.

24

SEVEN

WE DRIFTED FARTHER AND FARTHER AWAY. Our planet had become a small, pale blue marble, a dot we could barely distinguish in the distance. I felt a bit of homesickness and regret, but time passed, and it was remarkable how quickly it became normal again to be in space.

"You know what I'm thinking, sitting here?" Solon said one day. It was strange to speak of a day, of course, since there was no difference between day and night. We were no longer on a planet, where the sun only occasionally reached us because of the planetary rotation. "We've now experienced six of the seven eras in our lives."

"Six of the seven?" Aris asked.

"How did you arrive at seven eras?" Ensis challenged. "There have been many more."

"That's what I thought too," I said. It was long ago, but I remembered at least ten eras at the start, most occurring during the first three minutes after the Big Bang.

"You can divide history in all kinds of different ways," said Solon. "And I've come up with a new format."

"Well, okay then," said Ensis not very enthusiastically. "Go ahead and enlighten us. We can't go anywhere anyway."

"We don't have to do this," said Solon. "If you're not interested."

"I'm interested," I said quickly.

"I would like to know it too," said Aris. "And Ensis too, really. She's just a little grumpy because she can't go back to Earth."

"Shut up!" said Ensis. "Go ahead, Solon. I'm all ears, so to speak."

"The first era," Solon said solemnly. "Creation! The moment of the Big Bang and the first moments after that. The beginning of space and time. The period when the foundations of matter were formed and when we were formed."

"Okay," I said. "That's possible, if you like a broad format."

"In any case, it's nice that you think our origins are worthy of an era," said Ensis.

"And then era number two," Solon continued. "I was thinking that's the time when the stars and galaxies came into being."

"Agreed," said Aris. "A dynamic and impressive era when the whole cosmos was structured, when the immeasurable large space grew and grew and was filled with billions of galaxies."

"Including ours," I said. I glanced again at our Milky Way galaxy that we could see better than ever from where we were, with the luminous band high above us, containing more than two hundred billion stars, small dots, infinitely far away. Our own star, the sun. The Earth, the moon.

"And era three?" Aris asked. "That must be the origin of life?"

"Indeed. That started the moment when life came into being," said Solon. "That was something completely new."

Solon was right. I remembered it as if it were yesterday. The time we spent in the water, the moment when the first life arose—a cell that could reproduce itself. The origin of multicellular organisms. What ingenuity and beauty the Creator had hidden in the wondrous beginnings of the universe! So many surprises were waiting to be unfolded, one by one. Living cells that organized and copied themselves. The origin of plants and animals, fishes and reptiles, birds and mammals. The hominids.

"And then era four, of course," Solon continued.

"Let's keep that for another time," Ensis said. "I've found all this very interesting, but I can't handle all this information at once."

"Yes, but we've just come to the most interesting part!" I said. "Right, Solon?"

Solon laughed. "Yes, era four. That began at the moment the Creator spoke to *Homo sapiens*, when he invited them to continue together with him, the moment that he offered them life."

"He certainly had a lot of patience," Aris said suddenly, "when you look back on it. He waited long enough, until *Homo sapiens* had come far enough. Then he came to them. I still don't understand how things could go so wrong."

"I find it even more incomprehensible that the Creator continued with this project afterward," I said. "That he came up with a plan to save those people. With Abraham and all that. Even though it didn't quite become what he'd hoped, I'm afraid."

"Yes, but then came era five!" said Solon. "The coming of the Son of the Creator as a human baby."

"An unparalleled twist," Aris admitted. "No one could've thought of something like that."

"Exactly," said Solon. "Jesus revealed what life with the Creator was meant to be."

"He turned everything upside down," Ensis said. "I thought that was really exciting to see. How he became friends with beggars and criminals, how he treated women as his equals and gave children a place of honor, and how he healed the sick. With him it was all about love and mercy."

"That's why it was so bizarre that people didn't accept him in the end," I said, "that it ended in his arrest and his death. I thought that it couldn't possibly be happening, that the Creator wasn't going to let it come to pass. The darkness when he hung on that pole, remember? I thought the Creator's project had failed completely."

"I think that the soldier was the only one who realized what was happening," said Aris. "After that earthquake just after he died. 'This was the Son of God,' he said."

"And after that he turned everything around again!" exclaimed Ensis. "When he rose from the dead."

"Yes," said Solon. "That was really astonishing. No one could've seen that coming."

"It was totally impossible," I said.

"But Jesus, he did that."

"He did it. It was as if he'd walked straight through an invisible wall. No one before him had ever done anything like that. He opened a totally new way. The rescue plan of the Creator had succeeded in a totally unexpected manner."

"And era six?" Aris asked.

"That is the era of Breath. It's crazy that we never saw him in real life while he was at work on earth."

"Exactly," said Ensis. "I don't understand it either. Breath, strange name too."

"I've thought about it," I said. "People have to breathe, of course, to be able to live. Their body has to exchange oxygen and carbon dioxide."

Ensis sniffed. "Tell me something I don't know already."

"I think it's something similar with Breath," I said. "His name says something about what he does. I think that Breath is the one who blows good things into people. That he somehow helps them recognize Jesus as the Creator's Son, they begin to live differently, they become more patient, more loving."

"And they blow away the wrong things, you mean?"

"Something like that," I said. "That they quit doing wrong things." And I was silent. What did I know? I was only a proton.

"I think Pro's right," said Solon. "I think that Jesus' movement would've fallen apart very quickly if Breath hadn't been there. Nothing would've been left of it. It must've been his work that the group of a hundred Jesus-followers grew so incredibly large. Just think about it, how many are there now, after two thousand years? A couple of billion. That's an increase of a factor of over ten million, right? And they've had a great impact on the course of history."

"Yes, indeed," said Ensis. Her voice was scornful. "Those wars they fought in the name of the Creator, and the people they burned at the stake. They even thought they were doing the Creator a favor. How often they chose the temptations of power, riches, and self-righteousness instead of the contrarian teaching of Jesus. How powerful is Breath then, I wonder?"

"You're right," Solon admitted. "So much has gone wrong. I don't doubt the power of Breath, but I don't think he would force people, just as the Creator never forced people to make the right choices. But we've seen other things as well, haven't we? How followers of Jesus set up institutions to care for the handicapped, all those hospitals and schools they started, and their fight to end slavery. There really have been people who followed Jesus, however perfectly or imperfectly, who put their ego aside, who went against popular opinion and tried to act in his spirit. That must've been the work of Breath. He must've inspired people who've shown something of the new world that Jesus talked about so often."

"The development of science," I recalled. "That was quite a thing too."

I thought about the moment when Nicolaus Copernicus and Galileo Galilei discovered that Earth was not the center of the universe but just a planet that orbited around the sun. We hadn't known them but had heard much about them. Inspired by their belief that they should investigate the whole creation, they came to their revolutionary discovery. They were followed by a long line of men and women who, time and again, made new scientific discoveries.

We fell silent. I felt a strange dejection. Looking back, much had been achieved and wonderful things had happened. Nevertheless things just didn't look the way the Creator had intended them to be. There were so many things that just weren't right. There were so many people who lived in poverty or in war zones, so many children who were hungry. It seemed that the current world was groaning under the way humankind lived. Threats of nuclear destruction made things even worse. Humans would never succeed in realizing a new world on their own. In fact, it seemed that things were getting worse instead of better. At times, humankind seemed to threaten to completely destroy its own world—the water, the atmosphere, the soil . . . everything was polluted. Human activity had changed the climate dramatically, and that caused ever greater problems. Human beings hadn't carried out much of the Creator's mandate to care for the earth. The great opponent of the Creator might have been defeated, but powerless he definitely was not.

• • •

"Hey, Solon," Aris said suddenly, "didn't you talk about seven eras? You've only mentioned six."

"Yes, fortunately," Solon said cheerfully. "We now live in era number six. Much has been achieved but the project isn't finished yet. Era seven is the period to come, and which we're still waiting for. It's the grand finale."

"The time when Jesus will return?" I said, suddenly hopeful again.

"Yes, do you still remember what John told us about that? He talked about a time when evil will no longer have a place on earth, there will no longer be death and war, and the Creator himself will live among the people."

I tried to picture it in my mind. I thought of how Womuntu and Maisha once walked with the Creator in that brief period of happiness. Would that time come back? Would the Creator really want to live on earth among the people, eat and drink with them, and sit with them around a campfire? "A lot of things would have to change first," I said.

"Oh, but it's going to happen," said Solon. "It is going to come about. Really, Pro. It's going to be better than anything we've ever experienced—an era of new, exciting discoveries. Of celebration and adventure. Of harmony and love between the Creator and his entire creation. I think all of this is going to surpass all that we've seen over the last fourteen billion years. The Creator likes surprises. Many wonders are still awaiting us, I think."

"Us?" I asked. "Do you think we're going to be part of this?"

The pale blue dot had become so small in the meantime that it was almost indistinguishable. We slowly drifted farther and farther toward the edge of the solar system.

"The Creator loves to recycle, Pro," said Solon. "You know that, don't you? He's promised that he will use all the good things from the old world in his new world. So I wouldn't worry about it. Just be patient a little longer."

AFTERWORD

Deborah Haarsma

Sometimes our view of God is too small. In *Dawn,* the atheist astronaut Gary sees God as merely a filler for gaps in our scientific knowledge. Even Christians sometimes picture God only in relation to Earth and forget how the Creator is governing the entire cosmos.

In the Bible, we see the big picture, the true story of the world, the saga of God's work in creation and redemption. Yet our imaginations are stretched even further as we ponder what God has created in the natural world. Scientific discoveries are shedding new light on the story of everything, beyond what the Bible's original audience could imagine. Just recently, on Christmas Day 2021, NASA launched the James Webb Space Telescope. It will study the earliest galaxies and planets around other stars, multiplying the range of our vision beyond what even the powerful Hubble Space Telescope has opened to us. There is so much of our universe yet to see.

At BioLogos, we seek to help people understand how Christ-centered faith and the best scientific evidence available can fit together. One resource we've developed and shared on our website is called "The Big Story," narrated by Rev. Leonard Vander Zee. It tells the grand biblical narrative combined with scientific knowledge in an imaginative style, and we hoped that it might inspire other creative works, like *Dawn,* by people passionate about both faith and science.

In *Dawn* we read a beautifully imaginative journey of a tiny particle through the big story. Along with Pro, we experience the Big Bang, the origin and evolution of life, and key events in salvation history. I'm delighted to see how my friends Cees Dekker and Gijsbert van den Brink have brought together their scientific and theological expertise with the literary talents of Corien Oranje to tell a story that is scientifically accurate and inspiringly imaginative while holding true to the core of our faith.

Too many people think that science and biblical faith are in conflict. As a Christian astronomer, I see them working hand in hand. I am motivated in my scientific work by my understanding of Scripture. I believe that God's creation is comprehensible and am driven to keep searching for truth using the best tools I have. Books like *Dawn* invite us to keep asking questions, reevaluating our interpretations of Scripture and the universe, and exploring how faith and science can move together harmoniously in the world of the Creator.

Dawn is a tale about a Creator who is full of surprises. At each stage, we wonder along with Pro about what the Creator is planning and might do next. As Pro learns, the Creator likes to recycle, working through material and processes he has already made. God chooses to use as his instruments subatomic particles, recycled stardust, and even—especially—weak, fallible human beings. The Creator is also patient, allowing his plans to unfold over dozens, hundreds, and billions of years, on time scales boggling to humans.

Most of all, we learn that the Creator is driven by love, that "he has given his heart to *Homo sapiens.*" Even after we rejected him, he didn't give up. The biggest surprise is the "reverse Big Bang": the Creator coming into his creation as a vulnerable human being. In Jesus we have absolute assurance that God wants to be with the people he has made, and we have hope that there are wonders still to be revealed.

Like Pro and his friends, we readers often worry how everything will work out, especially when the light seems dim. Yet God is not daunted by our questions, and God has revealed enough for us to understand his overarching purposes. During the recent pandemic, I sometimes found hope by looking up at the sky. The stars remind me that there is a Creator, a loving, powerful Person who is above and beyond our earthly circumstances. He has not been defeated by Covid-19, and his love is as vast as the universe. We are not on our own. The Creator waited billions of years for us, and he won't give up on us now. His invitation is open for us to know him and be part of his plan.

After reading the manuscript of *Dawn*, one fifteen-year-old friend of ours told us, "I love Jesus, and I love science. But I never saw so clearly before how the two fit together!" My hope for this book is that many more readers will be inspired to grow in appreciation for our Creator and his redeeming work, for the gifts of science, and for the ways science and faith can both enrich our understanding of the other. The tale of all that came to be, after all, is still unfolding.

A CONVERSATION AMONG THE AUTHORS

As an epilogue, the three authors discuss among themselves the process that led to Dawn, *and they share what motivated them to write the book.*

Cees: Five years ago, Gijsbert and I organized a series of monthly meetings with a group of scientists, theologians, and philosophers from all over our country for in-depth discussions on evolution and the Christian faith. The meetings gave us wonderful insights, and we actually did not see any fundamental conflicts between creation and evolution. At the same time, we meet lots of fellow believers in daily life who nevertheless experience this as a difficult topic, because the story told by science so often sounds quite different from the big story about God.

Gijsbert: Many people are quite ready to accept scientific findings, but they understandably find it difficult to relate them to how the Bible describes God's history with this world. That led Cees and me to the thought that it would be a good idea to sketch some kind of "big narrative," let us say *the grand story of life*. That narrative would describe the history of all that is, from God's first creation acts up to the world of today, including important insights from the sciences. To the extent that we can comprehend that overarching story from our limited perspective, of course. Quite a few of these "big histories" have been published lately, so apparently people feel the need for reflection about our place in the history of the universe. Hence, plenty of good reason to write such a book. Yet this posed an enormous challenge. For such a project, one not only needs a lot of expertise in sciences and theology, but also a person needs a particular talent for telling stories.

Cees: Exactly! And as we got talking, we got the idea to ask Corien. I had greatly enjoyed working with her on a children's book on creation and evolution (*Science Geek Sam and His Secret Logbook*, Lion Children's, 2017). From that project it was clear that Corien is terrific in explaining complex scientific and theological concepts in understandable language, and she was able to turn even that into a humorous and gripping story. It would be really great if she would agree to take this on.

Gijsbert: So Cees emailed Corien.

Corien: That must have been the summer of 2017. I was hiking a trail with Dick, my partner, when I received the email from Cees. At that time, I had no idea what Cees and Gijsbert had in mind. Perhaps a simple version of Gijsbert's book, *Reformed Theology and Evolutionary Theory* (Eerdmans, 2020)? To take that very complicated text (sorry, Gijsbert) and turn it into something that even I could understand? That seemed to be a nice challenge. But then I heard what the intention was: a *novel* in which we would tell the story of the Big Bang and all of the world's history, integrated with the great narrative of the Bible. That seemed a rather tough challenge to me. Instantly, I blurted out a long list of names of people who seemed to be so much more fit to take up such a challenge than me. Maybe you had already asked them, and they didn't have the time or the inclination?

Gijsbert: Yes, Corien, we didn't really want to tell you, but now that you push us, indeed we had asked every Tom, Dick, and

Harry, and when we couldn't find anyone, we thought, well, let's ask Corien; she will do it for sure. . . . No, not really, you were, as Cees already said, the first person that we instantly thought of.

Corien: Ha ha! But I couldn't really picture how we could tell such a story. A common friend of ours, René Fransen, had alerted me to a speech, "From Stardust to the New Jerusalem," that Leonard Vander Zee delivered at the 2015 BioLogos conference on Evolution and Christian Faith (https://biologos.org/resources/from-stardust-to-the-new-jerusalem-gospel-centered-preaching-in-an-evolving-universe). It impressed me how Vander Zee was able to tell the story of God and creation anew, in less than ten minutes, and in doing so include wonderful elements of science in his presentation. I found that impressive, and I have often rewatched that speech. Something like that would be tremendous, but it seemed to me to be a rather difficult task.

Gijsbert: Fortunately you were willing to sit down with us to discuss whether we could find a good format for the book. On the evening of a conference about my book where all three of us were present, we had a long discussion about the how and why of such a novel. It should offer a real *narrative*. Not a dry summary of scientific and biblical knowledge, but a story. After all, we need stories to be able to understand reality around us and to understand our modest place in this reality.

Corien: That is true. That notion reminded me of the Narnia books and the Space Trilogy by C. S. Lewis, stories in which the gospel comes to you in an entirely new way. As a youngster, these books were very helpful to me.

Cees: Exactly. An engaging history of everything: about the origin of the cosmos, evolution, the dawn of humankind with Adam, Jesus, everything! Retelling the great moments in history from a Christian perspective: of creation in all its aspects from Big Bang to evolution, the call and fall of humanity, the coming of Jesus, until now. One story of "how all of it might have happened."

Gijsbert: Of course, what we derive from the Bible is of a different order than our scientific knowledge, in that the Bible is not just a history book. But I do think it is extremely important to continue to relate those two narratives to each other. Otherwise one starts living in two worlds that are unconnected to each other. I think that the Christian faith, in fact, has everything to do with our planet and our cosmos, with the natural world, with their origin, history, and destiny.

Cees: Notably, the target audience should be very broad, not a narrow elite club of academically educated experts but your brother Peter, who is a gardener and an elder in his local church, your niece in grade eleven, and anyone who ponders about questions around faith, the Big Bang, and evolution. And last but not least, also your agnostic neighbor who has a broad interest in science, religion, and the meaning of it all. But how do you tell that big story? We sat around for quite a while to talk about the format and perspective that we could choose.

Corien: Yes, the perspective is important. A story becomes more personal and interesting when you don't choose the perspective of an all-knowing narrator, but instead show the events from the point of view of someone who experiences them personally. But obviously, that is rather tricky when it comes to a story that begins with the Big Bang and continues until the second coming of Christ. Could the perspective of an angel be an approach, I wondered?

Gijsbert: And then you came up with the idea of a proton as storyteller, Cees.

Cees: Yep, a proton. Obviously, we couldn't choose a human being as a protagonist because people weren't there from the very beginning, fourteen billion years ago at the Big Bang. Some time ago, I read the novel, *The Portrait* by Willem Jan Otten (Scribe, 2014) in which a painter's canvas was the narrating protagonist. Quite a strange concept, but after a few pages, one gets used to it. And so, a proton seemed to be a suitable narrator for our book, for protons originated about one second after the Big Bang, and they are almost infinitely stable and hence did last throughout the age of the universe.

Corien: When you mentioned the name of Willem Jan Otten, I was intimidated a bit, as he is such a celebrated prize-winning novelist. I know my limitations. I'm not a highbrow literary author, right? But the enthusiasm of both of you about the idea of a proton was infectious.

Gijsbert: Once again, this was not a project with high literary ambitions. In fact, it had to be a story in rather plain language.

Corien: The aim of this project was very appealing to me. A book in which we show in a very accessible way that science and the story of the Bible do not exclude each other. That as a Christian, one does not have to be afraid of the discoveries of science. So I decided to take on the challenge.

Cees: Yes, we were very happy about that. As a team of three authors, we complement each other perfectly. Gijsbert as a theologian who did much thinking about the relationship between faith and science, and evolution in particular. I, as a scientist who is active in physics and biology (and in the distant past also trained as an astronomer). And you, Corien, as an experienced author with a background in theology. You would write the story, and we were going to give feedback, for example by providing and checking scientific information.

Corien: How many words per day would I write—you asked, Cees. I said, "Well, about a thousand on a good day."

"Ah, then you'll be finished in a month," you said. Ha ha! How long have we been working on it?

Cees: Two and a half years, I think.

Corien: Let me get my calculator for a moment. That's about forty-one words per day. Well, yes, this is not, say, a book for teenagers about basketball, where you can just continue to write when you have the inspiration. Inspiration was not enough. This is a book about topics that are enormously complicated: astronomy, geology, biology, and last but not least, hermeneutics. A giant undertaking. Initially, I had to do a lot of reading. Gijsbert's book, of course. Then literature about the origins of the universe, and then about geology and evolution. Fortunately, I was always able to come to you when it came to needing some help on these topics.

Cees: We had a *lot* of contact via zoom and email. I just looked it up: we exchanged 2,067 emails . . .

Corien: The consultations between the three of us were enormously helpful. We developed the story line together, and after each discussion, I could continue again. Sometimes I asked myself, When do these guys sleep? After each late-night question, I would promptly see the following morning that you had already sent me answers by email. For me as a writer, book writing is my full-time job, but for you, all this came in addition to your regular work at the university, of course. All in all, it was quite a job, also for the two of you.

Gijsbert: Granted. But this was all-important. The challenge was to communicate scientific insights in such a way that they do not alienate, but in fact, contribute to a genuine amazement about God's greatness. As the psalmist exclaims: "How great is our

Lord! . . . His understanding is beyond comprehension!" I think you've really succeeded in conveying a sense of that, Corien!

Corien: I hope so. Very frequently, I despaired during the writing process. I didn't seem to make any headway. How many editions of the first chapters did we have? I think it must have been more than fifteen. Still, I'm happy we have done this. For me, it was a great voyage of discovery. When you realize all the things that happened in the first *second* after the Big Bang. The many possible ways that things could have gone wrong, then and later in the history of the universe. It is a super-interesting topic!

Gijsbert: At the end, too, you came up with so many lovely stylistic flourishes. For example, when you hint that the eating of salami and other meats will soon be perceived as somewhat controversial and up for debate.

Corien: That was an idea from one of you, I think! Anyway, I thought that the writing would get easier once we arrived at the biblical story. However, we still had quite a few struggles with that. Especially with the piece about God's call to the human beings in Eden.

Cees: That was a central point for me. I found it enormously fascinating to jointly think how to characterize the key points of the Genesis text—the calling of humankind, the tasks given to humankind, the fall of the human pair that led to a separation from God, and so on—and how we could incorporate these elements into the format of the story.

Gijsbert: Yes, how do you incorporate the first chapters of Genesis? We view them as the Word of God. How do you take this text seriously in its full theological meaning without literally copying all sorts of aspects that have a symbolic meaning? At a certain point we decided to give Adam and Eve different names, however with comparable meanings, "human" and "life."

Cees: The origins of human beings lie in Africa, so that is where this story had to occur. So, they were given names from African languages. Similarly at some other spots in the narrative, we incorporated new images to express the core message of the story in new ways.

Gijsbert: The reader who is accustomed to the actual biblical story may feel a bit unsettled when reading these chapters in the book. At least, that is how I experienced it myself in cowriting them. But I think that our narrative does justice to the distinctive nature of the first chapters of Genesis, which are not merely historical accounts. They are not separate from history either, though. Especially on these chapters we exchanged a *lot* of emails, back and forth on every sentence, commas and periods included, so to speak. All three of us were aware of how important it was to listen to the text and put things properly into the words of the story.

Cees: I think that our wording of the paradise story may take some getting used to for many people. When you read a story a certain way all your life, and then suddenly you hear it told in a different way, then it may take some time to make that switch. But it may not just be Christians who are unsettled by *Dawn*. Some secular scientists may find it remarkable that *Dawn* sketches the hand of God in the origin of the cosmos and the development of life. On the other hand, it is of course also not entirely unexpected that Christians believe in a God who created and sustains this world.

Corien: The chapter about the birth of Jesus is also one we talked a lot about. How do we put words to the miracle that God became a human being?

Gijsbert: Yes, here we had to carefully consider how we could bring Proton's observations in agreement with what the church through the ages has confessed about God becoming human.

Cees: These are immense events that you put into words: Eden, the incarnation, Jesus' death and resurrection . . . Yet, this was an important, and mind-boggling exercise, because all this happened in the space and time of our physical world. Writing a story like *Dawn* forces one to imagine it all very concretely.

Gijsbert: *Dawn* outlines a way in which Christian faith and cosmic and biological evolution possibly can be thought of together. It does not make the claim that this is *the* way in which it happened. It sketches a possible scenario. We wrote this book in the way in which numerous children's Bibles have been written: as a retelling of the biblical story in a pictorial way, complementing all kinds of things and, of course, leaving out numerous details. In a similar approach, our book tells the grand story of God to adults.

Corien: We continued to edit until the very end. Just before the deadline, we still changed something in the first chapters because you thought that there was something not quite right about the photons, correct Cees?

Cees: Yes, that is right. In broad strokes, I am quite well informed about cosmology, but that is not the same as properly describing all the details in a book text. For example, in an earlier version, we described the occurrence of photons only in the recombination era, a couple of hundred thousand years after the Big Bang, when the universe became transparent to light. But, of course, photons originated much earlier, in the first moments after the Big Bang. Heino Falcke pointed that out to me after reading an early draft of the text.

Gijsbert: We benefited much from experts who read along with us in parts of the book. We want to express our great thanks to evolutionary biologist Duur Aanen, chemist Bauke Albada, astronomer Heino Falcke, science journalist René Fransen, geologist Nico Hardebol, theologian Almatine Leene, elementary particle physicist Frank Linde, astronomer Gergö Popping, theologian Eva van Urk-Coster, and developmental biologist Gert Jan Veenstra. They all gave valuable advice, for which we are very thankful. Of course, as authors we remain responsible for the final wording.

Corien: And we are grateful for all the great support from our Dutch Publisher, Royal Jongbloed, and now with Jon Boyd and all the staff at IVP, who made us feel welcome right from the start. And great thanks to the creative team at IVP—we love the cover design of *Dawn*! Finally, we are enormously grateful to our translator, Harry Cook, who, out of sheer enthusiasm, had already started translating this book before it was clear that it would be published.

Cees: And then there are the people who support our book with those great commending quotes. Many thanks for the support!

Corien: I really hope that this book will help Christians move forward in their thinking about these topics. It is a road that I have also traveled myself, and I have been greatly helped by the unproblematic, relaxed, and faithful way in which the two of you spoke about science and creation.

Gijsbert: I also see the book as an attempt to explain, in contemporary language, what the Christian faith encompasses. Hopefully, people who do not know much about it, but yet are curious, will find *Dawn* helpful.

Cees: My hope is that *Dawn* will show more people, skeptics and devout believers alike, that faith and knowledge connect more closely than often thought, and that people will marvel at the majesty of creation and the power of the gospel.

DISCUSSION QUESTIONS

THE ORIGIN OF THE UNIVERSE

1. Has your view of the origin of the universe changed over the years? In what ways? What has shaped your thinking?
2. How important is it for your faith if the universe came into being in a different way than you might have thought before?
3. What did you experience when reading chapters one to four of *Dawn*? Why?
4. If scientific evidence indicates that the universe is very old, how do you deal with the fact that the Bible seems to suggest a much younger age?

THE EVOLUTION OF LIFE

1. What is creation, and what do Christians confess when they say that God is the Creator of heaven and earth?
2. What do you think of the way Proton describes the origin of the first cell?
3. What do you think about the theory of evolution? What does it mean for your faith?

4. In *Dawn* there is already suffering and death before the fall of humankind. From evolutionary science, there are good reasons to think this was the case. How do you respond to this idea?

ADAM AND EVE

1. In retellings of the Genesis story (childhood Bibles, Sunday school stories, sermons, etc.), details are often included that are not in the biblical story itself. For example, the idea of Eden as a paradise in which all animals interact peacefully with each other is not explicitly stated in Genesis 1 or 2. How has hearing such stories influenced the way you have read Genesis 1–3?
2. How would you describe the uniqueness of humanity?
3. What do you think of the idea that Adam and Eve were not the first two people but rather chieftains or representatives of the earliest humans?
4. What do you think is the deepest core of Adam and Eve's sin?

RESTART

1. What stories would you have chosen if you wanted to summarize the Old Testament?
2. Which Old Testament story or character stands out to you? Why?

WHO IS JESUS?

1. What do you think of the way *Dawn* describes the incarnation as an "inverted Big Bang"?
2. What elements of Jesus' story particularly appeal to you?

3. Why is the resurrection of Jesus crucial to the Christian faith?

SCIENCE AND FAITH

1. Have you ever made a conscious choice whether or not to believe in God? What led you to it?
2. Do you think science and faith are at odds with each other, or do you think they complement each other?
3. What questions make it difficult for people to believe in the God of the Bible? To what extent does science play a role in this?
4. What questions or doubts do you have related to science and faith? What next steps can you take to address them?

BioLogos Books on Science and Christianity

BioLogos invites the church and the world to see the harmony between science and biblical faith as they present an evolutionary understanding of God's creation. BioLogos Books on Science and Christianity, a partnership between BioLogos and IVP Academic, aims to advance this mission by publishing a range of titles from scholarly monographs to textbooks to personal stories.

The books in this series will have wide appeal among Christian audiences, from nonspecialists to scholars in the field. While the authors address a range of topics on science and faith, they support the view of evolutionary creation, which sees evolution as our current best scientific description of how God brought about the diversity of life on earth. The series authors are faithful Christians and leading scholars in their fields.

www.ivpress.com/academic

biologos.org

MORE BIOLOGOS BOOKS ON SCIENCE AND CHRISTIANITY

How I Changed My Mind About Evolution
978-0-8308-5290-1

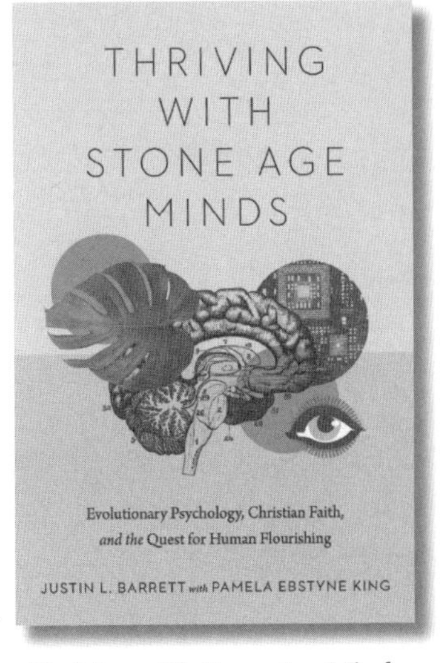

Thriving with Stone Age Minds
978-0-8308-5293-2

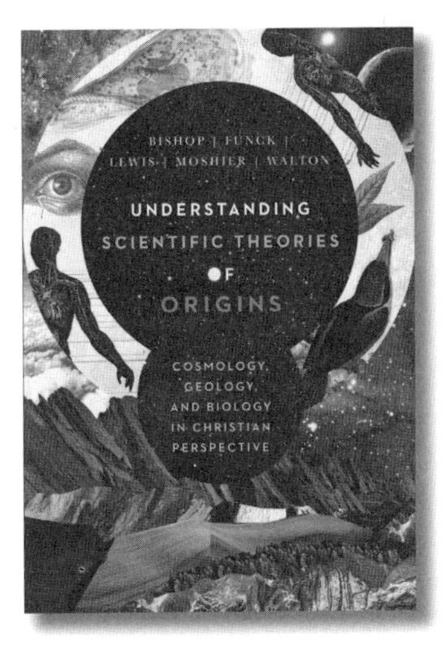

Understanding Scientific Theories of Origins
978-0-8308-5291-8